DICTIONARY OF FUEL TECHNOLOGY

DICTIONARY
OF FUEL
TECHNOLOGY

by

ALAN GILPIN, B.Sc.(Econ.), M.Inst.F., M.I.P.H.E.,
Assoc.I.E.Aust.

Chartered Fuel Technologist; Director of Air Pollution Control,
Queensland, Australia

PHILOSOPHICAL LIBRARY INC.

15 EAST FORTIETH STREET,
NEW YORK 16, N.Y.

Published 1969 by Philosophical Library Inc.
15 East 40th Street, New York 16, N.Y.
All rights reserved

*Printed in Great Britain for Philosophical Library
by Adlard & Son Ltd., Dorking, Surrey*

To

ELIZABETH, STUART AND DAVID

PREFACE

In writing this dictionary, the growing and dynamic subject of fuel technology has been treated broadly with the intention that the work should prove of interest and value to fuel technologists, chemical engineers, air pollution control engineers, economists, students and laymen alike.

Solid, liquid and gaseous fuels and other sources of energy, together with fuel-burning installations including nuclear reactors, are considered. Relevant scientific units of measurement, economic and commercial terms and organizations concerned with fuel and power are also included, and the important aspect of air pollution control receives special attention.

The work concludes with a guide to the principal books referred to in its preparation, summaries of conversion factors and useful data, and a temperature conversion table. Britain is now moving towards the International System of Units (SI) and, to assist readers with conversions from British and other non-SI units to the new units, conversion factors have been given in Appendix 2. It has not been considered desirable to place systematically, SI equivalents alongside every non-SI unit used in the Dictionary as in many cases space would permit approximate equivalents only while resulting in an undue cluttering of the text.

The dictionary is based largely on studies undertaken in the United Kingdom, and every effort has been made to include terms in use in the United States of America, Canada, Australia and New Zealand, and in Western Europe. Suggestions from readers for improving the range and value of the dictionary will be greatly appreciated.

A. G.

ACKNOWLEDGEMENTS

The author wishes to acknowledge with thanks the assistance of the following organizations who have kindly granted permission for the reproduction in this Dictionary of the diagrams indicated:

Her Majesty's Stationery Office: Figs. 5, 10, 15, 24, 28 from *The Efficient Use of Fuel*, Second Edition, 1958.

Central Electricity Generating Board: Figs. 7, 16.

National Coal Board: Figs. 2, 3, 9, 21 from the *Handbook on Mechanical Stokers for Shell Boilers*, 1952.

British Coal Utilisation Research Association: Fig. 30 from the *BCURA Annual Report*, 1965, p. 52; Figs. 4, 6, 32 from *Combustion of Pulverised Coal*, 1967, by Field, M. A., Gill, D. W., Morgan, B. B., Hawksley, P. G. W., published by BCURA, Leatherhead, Surrey, England.

British Standards Institution: Fig. 26 from B.S.2811: 1958.

The Institute of Fuel: Fig. 25 from *Developments in the Application of Two-Stage Combustion*, by Wheeler, W. H., Paper 7, Third Liquid Fuels Conference: Applications of Liquid Fuels, Torquay, England, 1966.

Leslie Hartridge Ltd., Buckingham, England: Fig. 12.

How Group Ltd., Stone, Staffordshire, England: Figs. 13, 22.

Tekni Gas Ltd., East Grinstead, Sussex, England: Fig. 17.

John D. Troup Ltd., London, England: Figs. 19, 31 from *Some Fundamental Aspects of Combustion in Fuel Beds*, by Gunn, D. C.

Engineering and Boiler House Review, Engineering Review Publishing Co. Ltd., Maidenhead, England: Figs. 1, 20, 27 from 'Oil Burners for Steam Raising Boilers', Nov. 1963, by MacCarthy, J. P. and Heitzman, P.F.

Industrial and Process Heating, Factory Publications Ltd., London, England: Fig. 29.

ABBREVIATIONS

absolute	abs	grain	gr
alternating current	a.c.	gramme	g
ampere	A		
ångström	Å	henry	H
atmosphere	atm	hertz	Hz
atomic number	at.no.	horsepower	hp
atomic weight	at.wt.	hour	h
		inch	in
boiling point	b.p.	inch of mercury	inHg
British thermal unit	Btu		
		joule	J
calorie	cal		
candela	cd	kilocalorie	kcal
centigrade heat unit	Chu	kilogramme	kg
centimetre	cm	kilogramme per cubic	
centipoise	cP	metre	kg/m³
centistokes	cSt	kilometre	km
coulomb	C	kilowatt hour	kWh
cubic inch	in³		
cubic centimetre	cm³	litre	l
cubic decimetre	dm³	lumen	lm
cubic metre	m³	lux	lx
cubic millimetre	mm³		
curie	c	maximum continuous	
		rating	m.c.r.
day	d	mega	M
dyne	dyn	melting point	m.p.
degree Celsius		metre	m
(Centigrade)	°C	metre per second	m/s
degree Kelvin	°K	metre per second	
degree Fahrenheit	°F	squared	m/s²
degree Rankine	°R	micron	μ
direct current	d.c.	milligramme	mg
		millilitre	ml
electromotive force	e.m.f.	millimetre	mm
		millimetre of mercury	mmHg
farad	F	minute	min
foot	ft	molecular weight	mol.wt.

xi

newton	N	square inch	in^2
newton per square		square centimetre	cm^2
metre	N/m^2	square decimetre	dm^2
normal cubic metre	Nm^3	square kilometre	km^2
normal temperature		square metre	m^2
and pressure	n.t.p.	square millimetre	mm^2
		standard temperature	
ohm	Ω	and pressure	s.t.p.
		stokes	St
poise	P		
pound	lb	tesla	T
pound-force	lbf		
poundal	pdl	volt	V

revolution per		water gauge	w.g.
minute	rev/min	watt	W
		weber	Wb
second	s	weight	wt.
second per second	s^2		
specific gravity	sp.gr.		
specific heat	sp.ht.	yard	yd

A

°**A.** Absolute temperature, in degrees Kelvin. The symbol ° K is now recommended in preference to °A. See **Kelvin Scale.**

Å. See **Ångström.**

Abel Flash Point Apparatus. Apparatus for determining the *flash point*, q.v., of petroleum products which flash below 120° F (49° C). It consists of a test cup to hold the sample, a lid which carries a thermometer, a test flame device and a stirrer; the test cup is mounted in a water-bath with an annular air space between the two. The temperature of the water-bath is slowly raised and with it the temperature of the oil sample; at regular intervals the test flame is applied through a special aperture to the interior of the cup, the temperature at which a flash occurs in the vapour space being recorded as the "flash point". See **Pensky-Martens Flash Point Apparatus.**

Absolute Humidity. The ratio of the mass of water vapour to the volume occupied by a mixture of water vapour and dry air. It is calculated:

$$d_v = \frac{m_v}{V} \text{ g/m}^3$$

in which:

d_v = density of water vapour;
m_v = mass of water vapour, g;
V = volume of water vapour and dry air, m^3.

See **Relative Humidity.**

Absolute Pressure. A pressure of −14·7 lb/in^2 or −760 mmHg. Thus the zero mark on a steam pressure gauge actually means a pressure of " one atmosphere " or 14·7 lb/in^2 abs. If a gauge indicates a pressure of 100 lb/in^2 this is the equivalent of 114·7 lb/in^2 abs. Thus absolute pressure = gauge pressure + 14·7 lb/in^2.

Absolute Temperature. Temperature measured from *absolute zero*, q.v.

Absolute Viscosity. See **Dynamic Viscosity; Viscosity.**

Absolute Zero. The temperature at which a *perfect gas*, q.v., kept at constant volume, exerts no pressure. It is equal to 0°R; 0° K; −460° F; −273° C. See **Temperature Scales.**

Absorption. The passing of a substance or force into the body of another substance; a liquid may be absorbed and held by cohesion

1

or capillary action in the pores of a solid; or gaseous molecules may be held between the molecules of a liquid. The characteristic waves of heat or light radiations may be retained by a solid, liquid or gas, being transformed into either kinetic energy or greater molecular vibrations, when the temperature of the absorbing substance rises; or into excited atoms or molecules when the substance becomes fluorescent. See **Adsorption**.

Absorption Coefficient. (a) the volume of gas, measured at normal temperature and pressure, dissolved by unit volume of a liquid under a pressure of one atmosphere. (b) The degree to which a substance will absorb radiant energy. When a parallel beam of radiation passes through a small thickness of x of a uniform substance, the fraction absorbed is $u_a x$, where u_a is the absorption coefficient of the substance for the radiation.

Absorption Control. In respect of a *nuclear reactor*, q.v., the use of a neutron absorber to absorb some of the neutrons and vary the reactivity. The absorber, usually cadmium or boron, is incorporated in control rods. The position of these rods can be varied in relation to the core of the reactor.

Absorption Oil. An oil of high affinity for the light hydrocarbons. See **Absorption Plant**.

Absorption Plant. A plant for recovering the condensable portion of natural or plant gas, by absorbing these hydrocarbons in an *absorption oil*, q.v., usually under pressure, followed by separation and fractionation of the absorbed material. See **Natural Gasoline**.

Absorptivity. The fraction of incident radiation which is absorbed by a surface on which it falls; a perfect absorber is a *black body*, q.v. The absorptivity of a material is numerically equal to its *emissivity*, q.v.

Abundance Ratio. The ratio of the number of atoms of one isotope to others of the same element, in a natural or enriched material. See **Isotopes**.

Accelerator. A machine for accelerating to high kinetic energy charged atomic particles such as electrons, protons, deuterons and helium ions. Common machines are the cyclotron, synchrocyclotron, synchrotron, betatron, linear accelerator and the van de Graaf accelerator.

Acetylene. C_2H_2. A colourless, poisonous, gaseous fuel. It may be produced by the chemical reaction of calcium carbide and water:

$$CaC_2 + 2H_2O \rightarrow C_2H_2 + Ca(OH)_2$$

Hydrated lime is a by-product of the process. It is also manufactured from methane, heavy gas oil and naphtha. In one example, oxygen and *natural gas*, q.v., are preheated to 950 to 1200° F (510 to 650° C) and mixed, there being insufficient oxygen for complete combustion; the reaction achieved is:

$$2CH_4 \rightarrow C_2H_2 + 3H_2$$

Acetylene is an important industrial gas; an oxygen and acetylene mixture (used with an oxyacetylene torch) burns with the highest temperature of any combustible gas and hence has great value in welding and the cutting of steel and other metals. The temperature of the inner brilliant part of the flame has been estimated in the order of 6000° F (3300° C). Acetylene has a gross calorific value of about 1500 Btu/ft^3. See **Neutral Flame.**

Acid Chimney. A chimney in which the temperature of the process waste gases is below the *dew point*, q.v., of the vapours so that acidic condensates are formed. Such chimneys require special acid resisting linings. In cases where the problem is marginal, insulation of the chimney may be sufficient to keep the gases above acid dew point.

Acid Cleaning. A common method of internal cleaning of pipework, tanks, heater shells, boilers and condensers; acid is circulated to remove dirt or scale.

Acid Refractories. Refractory materials containing over 90 per cent of *silica*, q.v.; they are used in open-hearth and other metallurgical furnaces to resist high temperatures and attack by acid slags.

Acid Sludge. A black, viscous residue left after the treatment of petroleum oils with sulphuric acid for the removal of impurities. The sludge contains both spent acid and the impurities removed from the oil. See **Acid Treatment.**

Acid Soot. Or acid smut; an agglomeration of carbon particles held together by moisture which has become acidic through combination with sulphur trioxide; soot particles range in size from $\frac{1}{32}$ in. to about $\frac{1}{8}$ in. in diameter. The carbon particles are mainly coke spheres produced during combustion. Acid soot emitted from chimneys leaves brown stains on materials and damages paintwork; the brown stain is caused by iron sulphate. The problem is mainly associated with oil-fired installations equipped with metal chimneys. The potential hazard can be reduced by using fuels of relatively low sulphur content; operating plant with a minimum of excess air in order to reduce the formation of sulphur trioxide; elimination of

3

air inleakage; the raising of back-end temperatures; the use of additives; the insulation of chimneys and ductwork; or by a combination of such measures. See **Dew Point.**

Acid Treatment. An oil refinery process in which unfinished petroleum products, such as gasoline, kerosine, diesel fuel and lubricating stocks, are brought into contact with sulphuric acid to improve their colour, odour, and other properties. See **Acid Sludge.**

Activity. The number of disintegrations per unit time taking place in a radioactive specimen. The unit of activity is the *curie*, q.v.

Additives. See **Detergent; Diesel Index; Gasoline Additives; Lead Susceptibility; Pour Point Depressant; Viscosity Index Improver.**

Adiabatic. Without loss or gain of heat to a system. Thus an adiabatic change is a change in the volume and pressure of a parcel of gas without exchange of heat between the parcel and its surroundings.

Adiabatic Efficiency. In respect of a *steam engine*, q.v., or *steam turbine*, q.v., the ratio of the work done per pound of steam to the available energy represented by the *adiabatic heat drop*, q.v.

Adiabatic Heat Drop. In respect of a *steam-engine*, q.v., or *steam turbine*, q.v., the heat energy released and theoretically capable of transformation into mechanical work during the *adiabatic*, q.v., expansion of unit weight of steam.

Adiabatic Lapse Rate. Scc **Lapse Rate.**

Adsorption. The taking up of one substance on the surface of another; adhesion. Adsorbents in industrial use include activated carbon, activated alumina, silica gel and Fuller's earth. See **Absorption.**

Advanced Gas-cooled Reactor (A.G.R.). An improved design of *nuclear reactor*, q.v., compared with its predecessor of Magnox design. As in the Magnox reactor, the heat of fission is removed from the reacting core by circulating carbon dioxide under pressure through it and passing this coolant gas through a boiler; graphite is also used as a moderator. The fuel elements however are made not of metallic uranium in a magnesium can, but of uranium oxide in a stainless steel can. This enables the heat of fission to be removed at a higher temperature in the coolant gas, and higher steam temperatures and pressures in the boilers can be achieved. The reactor shell may be protected from excessive heat by the use of a "hot box" or heat exchanger, the gases only coming into contact with the shell after being in contact with the heat exchanger. *Dungeness 'B' nuclear power station*, q.v., is of A.G.R. design.

Adventitious Ash. Incombustible materials such as shale, clay, *pyrites*, q.v., *ankerite*, q.v., dirt from earthy or stony bands in coal seams, and fragments of stone from the roof or floor of a seam, which are found in *run-of-mine coal*, q.v. During mechanical cleaning, much of the adventitious material can be removed. See **Ash; Inherent Ash.**

Aerated Burner. A *gas burner*, q.v., in which the gas induces primary air immediately before the burner ports. See **Bunsen Burner.**

Aerated Test Burner. A test burner designed to provide a rapid appreciation of gas quality; it measures the amount of air required to give a stable, well-defined, blue inner cone exactly $\frac{3}{4}$ in. high at a gas pressure of 2 5 in. w.g. The *A.T.B. Number* is the extent to which the air inlet shutter is opened to give the standard size of inner cone.

Aerosol. A particle of solid or liquid matter of such small size that it can remain suspended in the atmosphere for a long period of time; aerosols diffuse light, and the larger particles settle out on horizontal surfaces or cling to vertical surfaces. All air contains aerosols, the larger particles above 5μ in size being filtered out in the nose or bronchia. The smaller particles below 5μ in size pass into the lungs; they may be expelled immediately or retained for varying periods of time. Aerosols are classified into smoke, fumes, dust and mists.

After-burner. A burner located in the exit gases from a combustion process, providing sufficient heat to destroy smoke and odours.

After Heat. In respect of a *nuclear reactor*, q.v., heat produced by the decay of the fission products in the fuel elements, after the reactor has been shut down.

Air. A mixture of gases which forms the atmosphere in which we live. The constituents are nitrogen and oxygen, together with very small amounts of inert gases such as argon, krypton, xenon, neon and helium. Carbon dioxide is also present to the extent of about 0.03 per cent, together with water vapour and traces of ammonia, organic matter, ozone and salts. The relative proportions of the

TABLE 1—ASSUMED COMPOSITION OF AIR

Constituent	Volume %	Weight %
Oxygen	20·9	23·2
Nitrogen	79·1	76·8

permanent constituents show no appreciable variations, with the exception of water vapour. Nitrogen is quite inert in the atmosphere, merely diluting the essential constituent oxygen which supports life and combustion. For fuel technology purposes the trace gases may be ignored and the composition of the atmosphere, by volume and by weight, assumed to be as in Table 1. See **Excess Air; Nitrogen; Oxygen; Primary Air; Secondary Air; Tertiary Air; Theoretical Air.**

Air Assisted Pressure Jet Burner. A type of *pressure jet burner*, q.v., in which low pressure air is utilized to assist atomization, provide directional stability and promote primary zone turbulence. See **Oil Burner.**

Air Blanketing. An accumulation of air in a heat exchanger or other vessel which impedes or prevents the transfer of heat.

Air Count. The determination of the amount of radioactivity in a standard volume of air; in one method the prescribed volume of air is drawn through a filter paper on which the radioactive solids are deposited.

Air Director. See **Air Register.**

Air-dried Coal. Coal exposed to the atmosphere in a dry well-ventilated place, protected against the weather, so that it loses by evaporation most of its free or surface moisture. Coal is analyzed in this condition giving the percentages of *inherent moisture, volatile matter, fixed carbon* and *ash*, qq.v., See **Free Moisture; Proximate Analysis; Ultimate Analysis.**

Air Filter. A device for removing particulate matter from an air stream. See **Electrostatic Filter; Roller Filter; Viscous Oil Filter; "Zig-Zag" Filter.**

Air Inleakage. The leakage of air through defective brickwork settings and joints into flues. One of the most important features of boiler maintenance is the detection and elimination of such leaks. Leaks can be readily detected by holding a lighted candle, taper or duck lamp close to cracks in the brickwork and to joints where the infiltration of cold air might occur. Cracks and joints may be caulked with asbestos rope and sealed with a mixture of tar and fireclay or cement; warped flue inspection doors should be replaced; damper slots should be sealed with a twin roller system or a special air-tight damper cover; brickwork generally should be treated with three coats of tar or limewash to seal the pores. Exclusion of unwanted cold air from the flues improves boiler efficiency.

Airlift. The use of air or other gas to transfer liquids or solids from one part of a plant to another.

Air Pollution. Substances present in the atmosphere in concentrations great enough to interfere directly or indirectly with man's comfort, safety or health, or with the full use or enjoyment of his property. Air pollutants may be classified in three basic groups: (a) coarse particles of solids or liquids; (b) small particles of solids or liquids; (c) gases or vapours.

Air Pollution Index. An arbitrarily derived mathematical combination of air pollutants which gives a single number intended to describe the ambient air quality. The Department of Air Pollution Control, New York City, has established an air pollution index based on the continuously measured amounts of *sulphur dioxide, carbon monoxide,* and *smoke,* qq v, in the general atmosphere Smoke is measured in terms of a "coefficient of haze" (COH). The index is calculated as follows:

$$10(SO_2 \text{ ppm}) + CO \text{ ppm} + 2(COH)$$

Experience has indicated an average value for this index of about 12·0; an index reading of 50 or over is considered adverse and a cause for alarm.

Air Preheater. A device the purpose of which is to transfer heat from the flue gases to the air fed to the furnace for combustion purposes. It is usually situated between the economizer and the chimney stack. The recovery of heat from the flue gases reduces the heat loss from the chimney and raises the overall efficiency of the plant. It is claimed that a fuel saving of 1·5 per cent is obtained for every 35° F (20° C) reduction of flue gas temperature. The raising of the flame temperature in the furnace increases the rate of heat transfer by radiation, and reduces the amount of excess air required. Preheated air is essential for the combustion of low grade fuel. The combustion process is generally accelerated resulting in a short hot flame, and the reduced likelihood of smoke emission. Both regenerative and recuperative air preheaters are in use in industry, the latter being the usual type for use with boilers. See **Howden-Ljungstrom Air Preheater; Recuperative Air preheater.**

Air Preheater Corrosion. The wasting of the metal of boiler air preheaters caused by the condensation of acid from the flue gases on exposed metal surfaces.

Air Register. A combustion air regulator for oil burners. It permits the amount of air admitted to the combustion chamber to be controlled, i.e. kept in the right proportions with the amount of oil being consumed; it enables a flame front to be held at a satisfactory distance from the burner tip; in addition it causes the air to mix thoroughly with the oil by giving it a rotational flow which encourages turbulence. Figure 1 shows details of a compound air register. Also known as an *air director*. See **Oil Burner.**

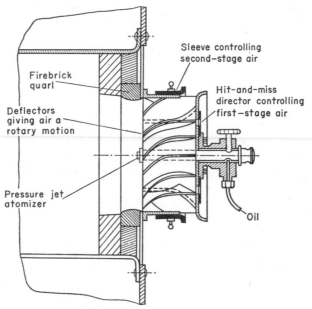

Fig. 1. *Compound air register.*

Ajax Furnace. A modified tilting *open-hearth furnace*, q.v., with provision for oxygen lancing; only during the charging, fettling and final phases of refining is fuel used, the necessary heat being supplied by the oxidation of the metalloids by jets of oxygen. Up to 1500 ft³ of oxygen per ton of steel may be used, production being much increased by this method. See **Steelmaking Furnace.**

Alcohol Fuel. Lower boiling point alcohols such as methyl and ethyl, which have high octane numbers (98 and 99 respectively), and are often blended with straight run gasolines to effect an upgrading of *octane number*, q.v. They may be added in a proportion of 10 to 20 per cent. Although these alcohols may be used alone in internal combustion engines, they are usually more costly and have much lower heat contents than ordinary *gasoline*, q.v.

Alkali, Etc., Works Regulation Act. See **Clean Air Legislation.**

Alkylation. An oil refinery process which plays an important part in the manufacture of motor gasoline and has a dominant role in making aviation gasoline. In the process of alkylation, butylene and isobutane are combined in the presence of sulphuric or hydrofluoric acid to produce a high percentage yield of aviation alkylate, which is largely iso-octane.

Alpha Particle. A positively charged particle emitted in the decay of some radioactive nuclei. It consists of two protons and two neutrons and is identical with the nucleus of the helium 4 atom. Alpha radiation has a power of penetration of less than one-tenth of a millimetre in human tissues; it is capable therefore of irradiating cells only in the immediate vicinity of the source of radiation.

Alternate Side or Wing-firing. A method of firing a boiler or furnace with solid fuel. In this method, sometimes known as the side-by-side firing method, the coal is thrown to one side of the grate only. This allows the volatiles released to be consumed by the hot gases from the other side. Once the coal has coked, the other side of the grate can then be charged. As the banked coal cokes it tends to fall into the centre. This system is excellent for avoiding the emission of smoke providing the little-and-often rule is strictly observed. There is less difficulty in keeping the grate properly covered with fuel, particularly at the back. This technique is more suitable for large and graded coals than for coal containing a high proportion of fines which tends to pack down, burning slowly and unevenly. See **Firing by Hand.**

Alternating Current (A.C.). An electric current the flow of which alternates in direction; the direction of flow of the electric current is reversed many times a second. The standard frequency in Britain is 50 cycles per second; in the United States 60 cycles per second. A cycle contains two reversals, namely, from one direction to the opposite, and then back to the original direction. The time flow in one direction is known as a half-period.

Alternator or Synchronous Generator. An alternating-current generator driven at a constant speed corresponding to the particular frequency of the electrical supply required from it. See **Turbo-alternator.**

Alumina. Al_2O_3. The trioxide of aluminium, an important constituent of ash; widespread in nature.

Aluminous Firebrick. A *refractory*, q.v., containing from 38 to 45 per cent of *alumina*, q.v., the balance being mainly silica.

Ambient Air Quality. A general term used to describe the state of the general atmosphere respecting the presence of air impurities. In Los Angeles, California, an attempt has been made to lay down ambient air standards as a guide to what concentrations of pollutants should not be exceeded, as a basis for public alerts. Most monitoring systems attempt to break down and list the impurities measured in air such as smoke, dust, sulphur dioxide, carbon monoxide, oxides of nitrogen and ozone. Some cities, among them New York, use their findings to construct an arbitrary index number. See **Air Pollution; Air Pollution Index; Los Angeles Smog.**

Americium. Am. An artificial *element*, q.v., at. no. 95, at. wt. 243. It is prepared artificially from *uranium*, q.v., in a cyclotron and is formed in very small quantities in nuclear reactors.

Ammeter. An instrument for measuring amperes, i.e., the current or rate of flow of electricity. See **Ampere.**

Ammonia. NH_3. A colourless, pungent gas with a boiling point of $-27°$ F ($-33°$ C), extremely soluble in water. It is obtained on a large scale by *ammonia synthesis*, q.v., and from the ammoniacal liquor produced in the carbonization of coal.

Ammonia Recovery. The recovery of *ammonia*, q.v., from coal gas produced during *carbonization*, q.v. Three methods of recovery are available;

(a) direct system, in which clean hot gases, to which additional ammonia from the distillation of ammoniacal liquor has been supplied, pass into saturators containing 77 per cent sulphuric acid saturated with ammonium sulphate, crystals of ammonium sulphate falling continuously to the base of the saturators from which they are removed and centrifuged;

(b) semi-direct system, in which the gas is cooled to near atmospheric temperature to remove much more ammonia in liquor form, following which the gas is reheated to 140° F to 176° F (60° C to 80° C) and acid extracted as in the direct system;

(c) indirect system, in which the gas is cooled to near atmospheric temperature to remove ammonia liquor and then extracted by water in scrubbers. The solution of ammonium salts is distilled with lime to recover ammonia vapour, this vapour being condensed to give ammonia solution (up to 25 per cent ammonia) or neutralized with sulphuric acid and recovered as ammonium sulphate.

Ammonia Synthesis. The manufacture of *ammonia*, q.v., from nitrogen in the air and hydrogen derived from *naptha*, q.v., and steam. In a typical process naphtha is vaporized and preheated in a refinery gas or naphtha fired heater and raised to a pressure of 470 lb/in² gauge. The naphtha vapour, together with some recycled hydrogen, is passed over zinc oxide and cobalt molybdate beds to remove sulphur compounds. Superheated steam is added to the vapour stream as it enters the primary reformer catalyst filled tubes; the tubes are heated by burners which use naphtha or refinery gas as fuel. The mixed vapour passing over the catalyst reforms into carbon dioxide (CO_2), carbon monoxide (CO), hydrogen (H_2) and methane (CH_4):

$$2C_7H_{15} + 14H_2O \rightarrow 14CO + 29H_2$$
$$CO + H_2O \quad \rightarrow \quad H_2 + CO_2$$
$$CO + 3H_2 \quad \rightarrow \quad CH_4 + H_2O$$

In the secondary reformer air is added to supply nitrogen (N_2). On passing over more catalyst methane is converted to carbon monoxide:

$$CH_4 + H_2O \rightarrow CO + 3H_2$$

The reformed gas passes into a CO converter where carbon monoxide and steam are converted into carbon dioxide and hydrogen:

$$CO + H_2O \rightarrow CO_2 + H_2$$

The gas is scrubbed in a tower with vetrocoke solution which absorbs most of the carbon dioxide; the vetrocoke solution is heated in a regeneration tower to liberate the carbon dioxide, which is vented to atmosphere. The remaining carbon oxides react with steam in a methanator to form methane. The gas is compressed in three stages to 350 atmospheres and passed into the ammonia synthesis circuit; as gas passes through the ammonia converter, a catalyst converts hydrogen and nitrogen to ammonia:

$$3H_2 + N_2 \rightarrow 2NH_3$$

On cooling, the ammonia condenses and is separated from unconverted gas in a catchpot; the ammonia liquid passes to a storage sphere while waste heat is used to raise steam.

Ampere. The unit of electric current; the ampere is the constant current which, if maintained in two straight parallel conductors of infinite length, of negligible circular sections, and placed one metre apart in a vacuum, will produce between these conductors a force equal to 2×10^{-7} newton per metre of length. As an indication of the order of magnitude of this unit, a 2 kW electric radiator on a 240 V circuit takes a current of 8·3 amperes. Named after the famous French physicist, André M. Ampère (1775–1836).

Analogue Computer. A *computer*, q.v., in which analogue representation is used, i.e. in which a variable is represented by a physical quantity, such as angular position or voltage, which is made proportional to the variable. The problems best solved by analogue computers are:

(a) dynamic problems involving fast rates of change of variables, which must be solved in real time;

(b) problems involving non-linearities, empirical relationships, etc. and requiring continuous solution. See **Analogue Controller.**

Analogue Controller. A device in which a measured signal representing a particular plant variable which has to be controlled is compared after amplification with a further signal representing the demanded value. The error signal formed is passed through a shaping network to produce the desired control equation, and then fed back to the system to actuate some device which corrects the original plant parameter. Examples include the temperature controllers on a nuclear power station and the use of analogue controllers for the run-up of large turbo/generator units. See **Analogue Computer.**

Aneroid Chamber. A type of draught gauge; it consists of a sealed metal chamber with flexible sides which expand or contract with changes in the difference between the pressure inside and outside the chamber, and a mechanism to magnify and record the movement of the chamber. For boiler house practice, instruments with a range of 1 in. pressure to 1 in. vacuum are available. See **Draught; Draught Gauge.**

Angle of Inclination. The natural angle which a solid fuel takes when piled; the angle will be increased as the moisture content increases. Examples are: coke, 45°; graded coal, 40°; washed small coal, 55°.

Ångström. A unit of length, symbol Å, equivalent to 10^{-8} cm (one hundred millionth of a centimetre). It is employed for measuring wavelengths of light.

Aniline Point. The temperature at which an oil first becomes completely miscible with an equal volume of aniline. A convenient laboratory test for assessing the proportion of aromatic hydrocarbons in hydrocarbon mixtures such as petroleum fuels; a low aniline point indicates a high aromatic content and a low *cetane number*, q.v.

Ankerite. A carbonate of calcium, magnesium and iron; it may be found as white partings within a coal seam although it is frequently associated with iron ores.

Annealing. A *heat treatment*, q.v., process; it is used to eliminate the effects of the cold working of metals by removing internal stress. It may also be used to improve electrical, magnetic, or other properties.

Anthracene Oil. A coal tar fraction, boiling above 518° F (270° C); in addition to anthracene it contains phenanthrene, chrysene, carbazole and other hydrocarbon oils.

Anthracite. The highest ranking coal to be produced by the physico-chemical alteration of peat; it is hard, dense and lustrous, does not break easily and is clean to handle. Difficult to ignite, it burns with a short intense flame and with the virtual absence of smoke. Anthracite contains between 5 to 9 per cent *volatile matter*, q.v., the calorific value is of the order of 14,000 Btu/lb. The best anthracites in the United Kingdom are mined in South Wales.

Anthracitization. The progressive conversion into *anthracite*, q.v., of plant remains which might have produced *bituminous coal*, q.v., under less severe geological influences.

Anthraxylon. A glossy constituent in coal formed from woody parts of plants, trunks, branches and twigs. *Vitrain* and *clarain*, q.v., are predominantly anthraxylon. A term used in the United States of America.

Anticyclone. In meteorology, a high pressure area with winds rotating clockwise around the centre in the northern hemisphere and anticlockwise in the southern hemisphere. There is a general slow descent of air over a wide area. Normally this subsiding air brings welcome fair weather because the cold upper atmosphere carries little water vapour and the air warms as it slowly flows ground-ward, superheating any vapour entrained. In summer, in the British Isles, an anticyclone generally means fine, warm, sunny weather. In winter however anticyclones bring dense sheets of strato-cumulus cloud giving typical winter gloom; as the air descends it is compressed and heated so that a deep inversion layer is formed which very often results in fog. See **Cyclone; Inversion; Subsidence Inversion.**

Antifoams. Usually polyoxide or polyamide compounds which, when added to boiler water, reduce or prevent the formation of foam. In consequence a higher level of dissolved solids content can be tolerated.

A.P.I. Gravity

A.P.I. Gravity. See Degrees A.P.I.

Apparent Specific Gravity. The *specific gravity*, q.v., of coal or coke inclusive of any voids within the pieces included in the sample subjected to test. See **Porosity; True Specific Gravity; Voidage.**

Aromatic or Benzene Series. C_nH_{2n-6}. Cyclic *hydrocarbons*, q.v., in which six carbon atoms are linked in a ring structure, each carbon atom being alternately doubly and singly linked with adjoining carbon atoms and joined to one hydrogen atom. The first member of this series is benzene, C_6H_6; others include toluene, $C_6H_5CH_3$, and xylene, $C_6H_4(CH_3)_2$. Aromatic rings can also possess side-chains, and two or more rings may be linked together through two common carbon atoms. Naphthalene, $C_{10}H_8$, consists of two benzene rings linked in this way; benzpyrene has five rings. Compared with the *paraffin series*, q.v., aromatic hydrocarbons have a greater tendency to cause smoke should the combustion conditions be incorrect. Aromatic compounds are chemically reactive, but show great thermal stability.

Arsenic. As. An *element*, q.v., in coal, at. no. 33, at. wt. 74·9216. The arsenic content of a coal seam may vary within wide limits over a relatively small area of a coalfield; the result of a survey of coals in the East Midlands of England suggested a range of 15 to 40 ppm As_2O_3.

As Received. A basis for reporting an analysis of coal; it includes the total moisture made up of *free moisture* and *inherent moisture*, *volatile matter*, *fixed carbon* and *ash*, qq.v. See **Proximate Analysis; Ultimate Analysis.**

Asbestos. An insulating material of varying grades; some grades are suitable for temperatures up to 1100° F (593° C). Asbestos lagging is available in rigid semicircular sections for pipes; asbestos composition is also available for boilers and may be readily applied. Steel or aluminium sheet may be used to protect the insulation.

Ash. The inert residue remaining when a fuel has been completely burnt. The ash content of coal may range from 1 to 30 per cent or more; in Britain 12 to 15 per cent are typical figures. If ash melts, the particles run together to form a molten mass which on cooling is known as *clinker*, q.v. Ash largely consists of silica and alumina. The ash content of liquid fuel is of considerable importance, as it is an indication of impurities which may cause wear in the fuel pumps and nozzles and, if they find their way into the cylinder, increased wear of cylinder liners and piston rings and pitting of exhaust valves and seats.

See **Extraneous Ash; Fly Ash.**

Ash Handling Plant. Plant for the removal of ash from a boiler house. Ash from a furnace accumulates in the boiler ash hoppers; high pressure water jets may be used to remove the ash from the hoppers and wash it through sluice ways into a receiving hopper or sump. In some plants, the ash is removed by a grab for loading into lorries or barges; in others the ash is passed through a comminutor and pumped to lagoons where it settles out. Ash from grit arresters may be removed by a suction and water system.

Ash Plate. See **Dumping Plate.**

A.S.T.M. Coal Classification. A system of coal classification devised by the American Society for Testing Materials. Under this system, coals containing less than 31 per cent *volatile matter*, q.v., on a mineral-matter-free basis are classified on the basis of *fixed carbon*, q.v., only; these are divided into five groups:

above 98 per cent fixed carbon⎫
   98–92 ,, ,,   ,,   ,,⎬ anthracites
   92–86 ,, ,,   ,,   ,,⎭
   86–78 ,, ,,   ,,   ,,⎫
   78–69 ,, ,,   ,,   ,,⎭ bituminous coals

The remaining bituminous coals, sub-bituminous coals and lignites are divided into groups as determined by the *calorific value*, q.v., of the coals containing their "natural bed moisture", there being eight groups in all with calorific values ranging from above 14,000 Btu/lb (7775 Chu/lb) to below 8300 Btu/lb (4610 Chu/lb). The classification also differentiates between consolidated and unconsolidated lignites, and between the weathering characteristics of sub-bituminous and lignitic coals. See **Coal Classification Systems.**

A.T.B. Number. See **Aerated Test Burner.**

Atmosphere, Standard. The atmosphere at "normal" pressure; one standard atmosphere, at 32° F (0° C), is equal to the following:

$$14 \cdot 695 \text{ lb/in}^2$$
$$2116 \text{ lb/ft}^2$$
$$1.033 \text{ kg/cm}^2$$
$$760 \text{ mmHg}$$
$$29 \cdot 92 \text{ inHg}$$
$$33 \cdot 9 \text{ ft } H_2O$$
$$407 \text{ in. } H_2O$$

The zero mark on a steam *pressure gauge*, q.v., means a pressure of one standard atmosphere and must be distinguished from *absolute zero*, q.v.

Atmospheric Temperature Inversion. See **Inversion.**

Atom. The smallest quantity of an *element*, q.v., which can take part in a chemical reaction; it has a diameter of the order of 10^{-8} cm and consists of a *nucleus*, q.v., of diameter 10^{-12} to 10^{-13} cm, around which electrons move in orbits.

Atomic Number. The number allocated to an *element*, q.v., when arranged with other elements in order of increasing atomic weight; it is equal to the number of protons in the *nucleus*, q.v., and, in the neutral atom, to the number of extra nuclear electrons.

Atomic Weight, Relative. A number that gives the mass of an atom of an element relative to that of the isotope of carbon, C^{12}, with an assigned atomic weight of 12. An early scale took hydrogen, with atomic weight 1, as the reference standard. From 1900 to 1961 oxygen was used as a standard of reference, with an assigned atomic weight of 16. In 1961 the International Committee on Atomic Weights decided that the isotope of carbon, C^{12}, be taken as the standard of reference, and that it be given the atomic weight 12. The replacement of oxygen as the reference changed the established atomic weights by about 0·004 per cent; the relative atomic weight of oxygen became 15·9994. The International Committee also decided that the term "atomic weight" should be changed to "relative atomic weight".

Atomization. The breaking into fine particles of a liquid fuel to ensure intimate mixing with combustion air, a prerequisite for efficient combustion. Some burners utilize burner tips of special design to atomize fuel oil supplied under pressure; atomizing pressures vary from about 250 lb/in² for light distillates, to 600 to 1000 lb/in² for heavy residual fuel oils. Other burners utilize air or steam as the atomizing medium. The normal size range of particles for good atomization is from 300μ down to 5μ. See **Oil Burner.**

Attemperator. Or de-superheater, a device to control any excessive superheat in a boiler; there are two main types. In the spray-type attemperator there is direct contact between the superheated steam and the cooling steam; in the surface-type attemperator the superheated steam passes through tubes around which the cooling steam circulates.

Attritus. A dull constituent of coal formed from finely divided materials, leaves, pollens, spores, seeds, resin, etc. A term used in the United States of America, synonymous with *durain* q.v.

Audibert-Arnu Test. A test for determining the coking properties of coal.

16

Auger Mining. A technique of mining employed in many strip mines where the overburden is too thick to permit economical continuous stripping; augers up to 60 in. in diameter are used to bore horizontal holes up to 200 ft in length into the exposed coal seams. The coal falls from the augers into a conveyor.

Authorized Fuels. Fuels which by The Smoke Control Areas (Authorized Fuels) Regulations, 1956, are suitable for use in smoke control areas. They include: (a) cokes of all kinds; (b) anthracite and low volatile steam coals; (c) briquetted fuels carbonized in the process of manufacture; (d) gas and electricity. In any proceedings in respect of the emission of smoke in a *smoke control area*, q.v., it is a defence to prove that the smoke was not caused by the use of any fuel other than an authorized fuel.

Automatic Boiler Control. A system which automatically controls the fuel and air supplies to a boiler, adjusting for the required steam output while simultaneously maintaining the best fuel/air ratio for high combustion efficiency.

Automatic Voltage Control. In respect of the *electrostatic precipitator*, q.v., a system designed to maintain optimum electrical conditions and high efficiency of collection under all boiler operating conditions. The system generally works by slowly increasing the voltage until the current, because of flash-over, exceeds a predetermined value; the voltage is then reduced a little and the cycle repeated.

Auxiliary-Fuel Firing Equipment. Fuel-burning equipment for supplying additional heat to a boiler or furnace to achieve higher temperatures than would otherwise be achieved for raising additional steam, increasing output or completing the combustion of combustible solids, vapours and gases.

Availability. Plant capacity available for use, usually expressed as a percentage of the maximum capacity. The percentage availability may refer to a particular time, such as a time of peak demand for steam or electricity, or may be the average taken over a period, for example a year. The availability factor is calculated:

$$\text{Availability factor} = \frac{\text{Hours operated or operable}}{\text{Total hours in period}}$$

See **Capacity.**

Aviation Gasoline. A gasoline similar to motor gasoline but having a slightly narrower boiling range, about 104 to 338° F (40 to 170° C), a lower vapour pressure, 7 lb/in^2 max. Octane rating and volatility are the two most important properties of these gasolines. Anti-knock properties are measured in a standard Co-operative Fuel Research engine. Two ratings are obtained; one simulates operation under take-off conditions (enriched fuel mixture) and the other simulates cruising conditions (weak fuel mixture). See **CFR Engine.**

Aviation Mixture Methods. Tests to determine the octane numbers of high octane aviation fuels. In the Aviation Lean Mixture Method an engine speed of 1200 rev/min is employed; in the Aviation Rich Mixture Method an engine speed of 1800 rev/min is employed. See **Octane Number; Performance Number.**

Avogadro Hypothesis. A statement that equal volumes of different gases at the same temperature and pressure contain the same number of molecules.

Avogadro Number. Or constant; the number of molecules in a gramme-molecule. It is equal to $6 \cdot 02252 \times 10^{23}$ mole^{-1}.

Axial Flow Fan. A type of fan in which the direction of flow of the air from inlet to outlet remains unchanged. See **Centrifugal Fan.**

B

Back-pressure Turbine. A *steam turbine*, q.v., in which exhaust steam is utilized for process work or for heating. Thus the steam is not condensed at the turbine, being utilized at the temperature and pressure at which it leaves the turbine. See **Pass-out Turbine.**

Backward-curved Fan Blading. A design of blading for *centrifugal fans*, q.v., which is considered one of the most suitable for supplying forced draught to a boiler; it has reserves of pressure with which to overcome dirty boiler conditions, that is, it permits a wide variation in delivery pressure for the same quantity of air delivered. Blading of this design offers a high efficiency, with a higher initial cost and lower running costs compared with *forward-curved fan blading*, q.v. See Figure 2. See **Fan.**

Baffle. A refractory construction whose function is to change the direction of flow of the products of combustion.

Bagasse. The fibrous portion of sugar-cane after the juice has been extracted. Mills crushing sugar-cane commonly use bagasse as fuel for steam production. In mills not used for the refining of sugar, the supply of bagasse equals and often exceeds plant steam demand; there is thus no incentive to burn bagasse efficiently, and

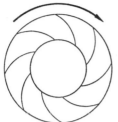

FIG. 2. *Backward-curved fan blading.*

steam boilers serve also as incinerators. A typical percentage analysis of bagasse, as fired, is: moisture, 52; volatile matter, 40; fixed carbon, 6; ash, 2; with a gross calorific value of 4000 Btu/lb.

Bag Filter. A device for removing particulate matter from the waste gases of industrial processes; the filter medium is a woven or felted fabric usually in the form of a tube. The bags may be up to 30 ft in length and up to 30 in. in diameter. The upper ends are closed, while their lower ends are connected to a gas inlet header which also serves as a hopper to catch the dust which is shaken down from the bags by mechanical or air-reversal methods. The collecting efficiencies of bag filters are high, between 99 and 99·9 per cent; low gas velocities of the order of 2·5 to 10 ft/min are required. The choice of materials for bag filters is limited by temperature limitations, being: (a) wool or felt, 200° F (93° C); (b) nylon, 400° F (204° C); (c) glass fibre, siliconized, 600 to 700° F (316 to 371° C); and (d) glass fibre, siliconized and graphite impregnated for longer life, 500° F (260° C). Pilot plants have been constructed at two British power stations to appraise the suitability of bag filters for power station emissions.

Bahco Centrifugal Dust Classifier. Apparatus for dividing samples of dust into size fractions by means of centrifugal force acting against an inward air current. By using different inward air velocities while the centrifugal force remains constant the sample can be divided into a number of size grades.

Bailey Wall. A wall of plain water tubes in the combustion chamber of a *water-tube boiler*, q.v., faced with metal blocks.

Balanced Draught. A combination of *forced draught*, q.v., and natural or *induced draught*, q.v., which ensures that the pressure of the atmosphere outside a furnace and the pressure of the gases inside (over the fire grate) are virtually the same; this implies zero on a water gauge. In practice, to avoid accidents, a slight suction is kept inside the furnace of about 0·05 in. w.g. See **Draught.**

Ball Mill. A low-speed *pulverizer*, q.v., for the production of *pulverized fuel*, q.v. It consists of a large diameter shell, tubular or conical in shape, containing a charge of steel balls about 1¼ in. in diameter. The shell is mounted on hollow trunnions which allow entry and exit for the dust-carrying air. The mill rotates at a speed of about 15 to 25 rev/min. The ground product passes through a classifier which separates the overlarge particles and returns them to the feed end of the mill. The outlet temperature aimed at is usually 120° F (49° C).

Ball-mill Method. A method for determining the *grindability index*, q.v., of a coal. A sample of coal is placed in a ball mill and the number of revolutions required to grind it so that 80 per cent of the sample passes a No. 200 U.S. sieve (74μ) is ascertained.

Ball-race Mill. A *pulverizer*, q.v., which grinds by crushing; a lower ball-race is turned by a motor causing the balls to rotate. Coal enters the centre of the mill and is thrown out against the moving balls which crush it against the races. The fine coal spills over the lower race where it is picked up by air and carried around the upper ring to the centrifugal classifier section. Particles of sufficient fineness are carried with the air stream to the burner; the oversize particles are returned for additional pulverization.

Ballast Gas. An inert gas introduced into a mixture of combustible gases to adjust its specific gravity.

Balloons, "Zero Weight". Special hydrogen-filled balloons, adjusted to "zero weight", used in air pollution research into the rise of chimney plumes under varying meteorological conditions. The balloons are introduced into a chimney and carried aloft into the general atmosphere by the chimney gases, their trajectory tending to follow the path of the chimney plume. A specially-mounted range finder can be used to plot the distance and the height of each balloon from time to time.

Band Screens. Rectangular perforated screen plates, connected together as a continuous band. They are used for screening circulating water in power stations.

Banking. The retention of a fire on a grate but at a very slow rate of combustion; this enables the boiler to be brought up to full steam quickly when steam is demanded. The banking of fires overnight or at the weekend is a common practice. The amount of fuel burned during the banked period should be only just sufficient to make up for natural cooling and to produce any small amount of steam that may be required.

Bar. The unit of atmospheric pressure, being equal to the pressure of one million dynes per square centimetre; the bar is equal to the pressure of 29·530 in. or 750·062 mm of mercury at 32° F (0° C).

Barn. An area of 10^{-24} cm²; a unit used to measure the cross-sections of a *nucleus*, q.v.

Barometric Condenser. A device which, by condensing steam, produces a partial vacuum in a piece of equipment; it is used in conjunction with oil refinery vacuum distillation towers. The barometric seal legs end in a hot well which contains both hydrocarbon and aqueous condensates. The hot well should be sealed and fitted with a vent whereby released vapours can be passed to a furnace and burned. The barometric water flows to an oil interceptor; the problem of odorous water may be largely solved by the application of odour-masking compounds.

Barometric Damper. A pivoted, balanced plate, normally situated in the flue between furnace and stack, and actuated by the draught. See **Damper.**

Barrel, United States. Normal measure used throughout the oil industry, the United States barrel equals 42 United States gallons or 35 Imperial gallons. Since the density of different crude oils varies, and the density of refined petroleum products varies widely, there is no universally applicable conversion factor from barrels to tons. However, for crude oils of a gravity of about 33 to 36 *degrees API*, q.v., corresponding to the averages for Iran, Iraq and Kuwait, there are roughly 7½ barrels to a long ton (2240 pounds). For refined products the conversion factors vary from about 6¾ barrels of heavy fuel oil per ton to about 9 barrels of aviation gasoline per ton. A useful rule of thumb in expressing oil supplies (e.g. rate of flow of wells, or the rate of exports of a given country) is that one barrel of crude oil a day represents fifty tons a year.

Barrels per Stream Day. The quantity of crude oil processed by a refinery in a continuous twenty-four hour period, expressed in barrels.

Barring Gear. Equipment for slowly rotating a turbine shaft when the turbine is off-load.

Base Exchange Process. See **Ion Exchange Process.**

Base Load. A steady demand equal to a high percentage of the installed capacity of a plant. See **Base Load Station.**

Base Load Station. A power station whose generating costs are

B   21

low and which stands, therefore, high in the "order of merit"; in consequence it is under load by day and by night for long periods. Nuclear power stations are by their nature base load stations, because of low generating costs and of a technical need for continuous operation at high load factors. See **Merit Order.**

Batch Process. A process for carrying out a reaction in which the reactants are fed in discrete and successive charges; not a continuous process. Examples are the *batch still, coke oven,* and the *horizontal retort,* qq.v.

Batch Still. A still in which distillation is carried out in "batches", the entire charge being introduced before heating begins, and the distillation being completed without the introduction of an additional charge; an intermittent or non-continuous process.

Battersea Gas Washing Process. A method of wet scrubbing the flue gases from the coal-fired Battersea power station, located in London on the Thames, to remove large quantities of sulphur dioxide. The washing medium comprises water from the Thames with the addition of chalk. About 35 tons of water are required for each ton of coal burnt. The efficiency of removal of sulphur dioxide is of the order of 95 per cent. The effluent is discharged into the river as an almost saturated solution of calcium sulphate. A similar system is used at Bankside power station, also situated in London on the Thames, where oil containing 3 to 4 per cent sulphur is consumed. The process has several drawbacks: (a) the plume is cooled and the residual sulphur dioxide and moisture fall readily on to the surrounding district; (b) the sulphur is not recovered in a useful form; (c) enormous quantities of water are required and water pollution occurs; (d) it promotes corrosion problems and adds significantly to generating costs.

Baum Coal Washer. A wet cleaning plant for separating "dirt" in the form of sandstone, shale, clay, pyrites, calcite and so on, from *run-of-mine coal,* q.v. Coal is passed over an immersed perforated grid; alternating pressure and suction cause the water to move up and down through the perforated bed on which the coal is travelling. The dirt settles while the clean coal rises to the upper layers.

Baumann Exhaust. A special design of the exhaust end of a turbine to give additional effective area for the exhaust steam to pass to the condenser.

Bed Moisture. The total moisture in a coal seam prior to working. See **Moisture.**

Belgian Grate-fired Kiln. An annular longitudinal arch kiln used in the brick-making industry; coal is fired on to grates situated near the bottom of the kiln. In some cases oil firing through the wickets is practised. See **Brick Kiln.**

Beneficiation. In relation to ores, a concentration process which may include drying to reduce the water content, roasting to reduce the sulphur content, and washing to remove some of the gangue or undesirable impurities.

Benzene. C_6H_6. A colourless, highly inflammable liquid; a member of the aromatic series of hydrocarbons. It is produced from *coal tar*, q.v., and coke oven gas. Benzene is also an old term for light petroleum distillates covering the gasoline and naphtha range.

Benzole. A product derived from the carbonization of coal and consisting essentially of *benzene*, q.v., together with its homologues, toluene and xylene; unsaturated hydrocarbons and sulphur compounds may also be present.

Benzpyrene. A hydrocarbon present in coal smoke and cigarette smoke; it is strongly suspected of being carcinogenic to man. See **Carcinogenic Compounds.**

Beryllium. Be. An *element*, q.v., and light metal, at. no. 4, at. wt. 9·0122. It is used in crystalline form for transistors and in fluorescent lighting, and in fabricated form for small aircraft parts. Its low atomic weight, low neutron-capture cross-section and high thermal conductivity render it suitable for use as a moderator and reflector in nuclear reactors; its considerable strength and high melting point make it a possible canning material for fuel elements. It is expensive, toxic and difficult to fabricate. See **Cross-section, Total.**

Bessemer Converter. A *steelmaking furnace*, q.v., consisting of a cylindrical body of steel plates, brick-lined, with a base provided with a series of air holes; the capacity of a converter varies from 10 to about 60 tons of metal. Molten pig iron is poured into the converter; air is forced through the liquid metal producing a flame at the mouth of the converter; no additional heat source is necessary. The tropena is a small side-blown converter. The scope of steelmaking by the Bessemer process has been widened by the adoption of an oxygen/steam blast. The process is then known as the V.L.N. (very low nitrogen) steelmaking process; by this process low nitrogen steels may be produced.

Beta Particle. An *electron*, or a *positron*, q.v., emitted in the decay of some radioactive nuclei. In beta radiation, the rays travel distances of up to a few millimetres in human soft tissues before they are fully absorbed.

BeV. A unit used in the United States of America equal to a billion electron volts; i.e. 10^9 electron volts. In Europe this is often denoted by GeV (G standing for giga).

Billet. A $3\frac{1}{2}$ in. to 5 in. square semi-finished steel shape, smaller than a *bloom*, *q.v.*, and usually rolled from a bloom.

Bimetallic Strip Thermometer. A device for measuring temperature. It comprises two strips of metal having different coefficients of expansion welded side by side; a temperature change causes the strips to curl or deflect slightly. The use of this device is usually limited to simpler types of thermostat; it is rarely used in industry for the straight forward measurement of temperature. See **Pyrometer; Thermometer.**

Bin and Feeder System. A system for supplying *pulverized fuel*, *q.v.*, to boilers or furnaces. It consists of a pulverizer of the slow-, medium- or high-speed type, from which pulverized coal is extracted and delivered by an exhauster fan to a cyclone situated above a bin; this cyclone extracts the coal dust which is deposited in the bin, at the base of which are a number of feeders, usually of the screw or rotary drum type, which distribute the pulverized fuel to the boilers or furnaces. The fuel falls from the feeder into an air stream from a primary air fan which conveys it through ducts to the burners.

Biological Shield. A thick wall or shield, usually consisting of 8 to 11 ft thickness of concrete, surrounding the core of a *nuclear reactor*, *q.v.*, to absorb neutrons and gamma radiation for the protection of operating personnel.

Birdnesting. The adhesion of fused ash to boiler tubes, cementing together unburned carbon.

Biscuit Firing. A stage in the firing of pottery ware in which the clay is transformed into fired ware at a temperature of about 2100° F (1150° C). See **Enamel Firing; Glost Firing.**

Bismuth. Bi. A grey-white metallic *element*, *q.v.*, at. no. 83, at. wt. 208·98. The metal is used as a constituent of fusible alloys. The naturally occurring isotope $_{83}Bi^{209}$ is the heaviest stable nuclide. The metal has a low neutron-capture cross-section and a low melting point of 520° F (271° C); it has been considered as a solvent for uranium in a liquid metal fuelled *nuclear reactor*, *q.v.*, and as a liquid carrier for fuel slurries.

Bitumen Oxidizing Plant. Plant in which air is blown through bitumen in order to convert it to a rubbery consistency which is required for special uses such as roofing felt and battery tops. The spent air from the blowing is malodorous because of incipient cracking and oxidation, and is piped to process furnaces where the odour is destroyed by combustion.

Bituminous Coal. The best known of solid fuels; the description "bituminous" was originally applied because of the tendency to burn with a smoky flame and melt when heated in the absence of air. The volatile content varies from about 20 to 35 per cent; the *fixed carbon*, q.v., between 45 and 65 per cent; the *inherent moisture*, q.v., between 3 and 10 per cent; and the calorific value between 11,000 and 13,800 Btu/lb (6110 and 7665 Chu/lb). Most bituminous coals have a banded or laminated structure, and a shiny black appearance. See **Coal.**

Black Body. Surfaces with an *absorptivity* and *emissivity*, q.v., of unity. The term "black body" does not necessarily mean that the body appears black to the eye. The radiation emitted by a black body depends upon its surface area and temperature and follows a relationship known as the *Stefan-Boltzmann law*, q.v.

Black Centre Burning. The condition of a fuel bed in which an area or areas of unignited coke appear at the surface. It may occur in underfeed stoking if the air rate is too great. Black centre burning may also occur due to the formation of "coke trees" from strongly coking coals. Black centre burning is accompanied by smoke emission. See **Red Top Burning.**

Black Smoke. Smoke produced when particles of carbon are derived from the cracking of hydrocarbon gases, following sudden cooling. Smoke is visible evidence of incomplete combustion. Black smoke is frequently considered to be smoke as dark as, or darker than, shade 4 on the *Ringlemann Chart*, q.v. See **Brown Smoke.**

Blast Atomizer. A type of *oil burner*, q.v., in which the atomizing medium may be air or steam. Blast atomizers are classified into three types according to the pressure of air or steam: (a) low pressure, operating with an air blast pressure of between 12 to 30 in. w.g. with about 25 to 40 per cent of the necessary combustion air used for atomization; (b) medium pressure, operating at an air blast pressure of between 3 to 15 lb/in^2 with about 3 to 5 per cent of the necessary combustion air used for atomization; (c) high pressure, operating with a steam or air blast pressure of 15 lb/in^2 and above.

Blast-furnace. A shaft furnace used for the reduction of ore to metal; used in the smelting of iron ore, copper, lead, antimony, tin, nickel and cobalt. In the steel industry, the furnace shaft is a vertical steel cylinder lined with firebrick standing up to 100 ft or more in height; the hearth diameter varies up to over 30 ft. Immediately above the hearth section, the diameter of the furnace increases to form a " bosh " which accommodates the *tuyeres*, q.v.; above the bosh the furnace diameter decreases gradually. The conical shape of the stack allows the charge of ore, coke and flux to swell; the bosh assists in checking descent. The charge is admitted through a double bell arrangement at the top. See **Blast-furnace Gas; Blast-furnace Stove.**

Blast-furnace Gas. A gas used extensively in the metallurgical industries, being produced in blast-furnaces during the reduction of iron ore. The calorific value of the gas varies, but a figure of 90 Btu/ft^3 (50 Chu/ft^3) is general. A typical analysis of the gas, the constituents being expressed in percentages, would be: carbon monoxide, 28; hydrogen, 2; carbon dioxide, 12; nitrogen, 58. The gas as it leaves the blast-furnace has a high dust burden; cleaning in two stages is commonly practised.

Blast-furnace Stove. A stove for supplying preheated air to a *blast-furnace*, q.v. A typical stove consists of a vertical steel cylinder with a spherical dome top; the stove contains a combustion chamber with a lens-shaped horizontal cross-section, while the major part of the volume of the stove is filled with *chequer-brickwork*, q.v. A stove passes through a two-stage cycle lasting perhaps 4 hours. In the first " on gas " stage the chequer-brickwork is heated by the combustion of blast-furnace gas for about 3 hours; in the second " on wind " stage the chequer-brickwork is cooled by the passage of cold air for about 1 hour. At first only part of the air to be heated passes through the stove, emerging at perhaps 2000° F (1093° C) when it is mixed with cold air to supply a blast at about 1000° F (538° C). As the chequer-brickwork cools an increasing proportion of the blast air passes through it; in this way the final temperature of the air supplied to the blast-furnace is maintained. With a battery of 4 stoves, a blast-furnace receives a constant supply of preheated air. Also known as a Cowper stove. See **Howden-Ljungstrom Air Preheater; Pebble Stove.**

Blast Saturation Temperature (B.S.T.). The criterion of the proportion of water vapour in the air-steam blast of a *gas producer*, q.v.; the maximum thermal efficiency of a gas producer is usually attained when the B.S.T. is about 140° F (60° C).

Blending. The mixing of two or more fuels of different properties and characteristics to produce a fuel with certain qualities; similarly, the mixing of various refined fractions to obtain lubricating oils having particular properties.

Block Heating. See **District Heating.**

Block Tariff. A method of charging for gas or electricity in which the price per therm or unit is highest for a specified initial number, or block, of therms or units consumed within a prescribed period, and lower for additional quantities of therms or units consumed within the same period. See **Tariff.**

Bloom. Semi-finished steel having a cross-sectional area greater than 36 square inches.

Blow-by. Gases consisting of carburetted gasoline together with some exhaust products which blow past the piston rings into the crankcase of an internal combustion engine. For many years American-made automobiles utilized a system of crankcase ventilation employing a road draft tube. With this system, air flowing past the moving vehicle aspirated crankcase fumes through the tube from the crankcase. Replacement air was drawn through a combination oil filter cap and air inlet. Devices to control crankcase emissions became mandatory in California from 1961 onwards. Control has consisted in returning crankcase ventilation air and gases to the engine intake manifold for consumption during operations. The negative pressure of the engine induction system is used to establish a positive flow of the crankcase ventilation air through the crankcase. This technique has solved between 20 and 40 per cent of the air pollution problems of petrol driven vehicles.

Blow-down. The release of water from the lowest part of a boiler to free it of sludge and reduce the dissolved solids content of the water. The amount of blow-down required per day depends upon the maximum total dissolved solids (T.D.S.) which can be tolerated without risk of *priming*, q.v. The number of gallons to be blown down each day may be calculated:

$$Y = \frac{ax}{b}$$

where, Y = amount of water to be blown down per day, gal;
a = total dissolved solids (T.D.S.) in feed water, gr/gal;
x = makeup water added per day, gal;
b = T.D.S. in blow-down, gr/gal.

Blow-down Valve. A valve connected to the lowest part of a boiler to fulfil two functions: (a) The boiler can be emptied for inspection and maintenance purposes; (b) the sediment can be removed and

27

the salt concentration in the boiler water reduced. In horizontal boilers, the valve is connected to the base of the boiler front plate. In water-tube boilers, it is connected to the mud drum by a substantial steel pipe. Many boilers are equipped with continuous blow-down valves through which a constant stream of water leaves the boiler. Without blow-down sediment would simply accumulate and the dissolved solids concentration increase until foaming and priming occurred, resulting in wet steam. Blow-down valves should be provided with a safety device to prevent the removal of the key until the valve is fully closed. The parallel slide valve is the best type that can be obtained.

Blow-holes. Holes which appear in a fuel bed through localized acceleration of combustion rates.

Blue Water Gas. See **Water Gas.**

Boghead Coal. A coal derived from *sapropel*, similar to *cannel coal*, q.v., but characterized by the presence of algal remains and substantial enrichment in hydrocarbon compounds. A variety of boghead coal known as torbanite is largely a compound of minute algal bodies, and yields over 90 per cent of volatile matter.

Boiler. A device for making steam from water, although the term is also applied to many hot water appliances in which the water is not meant to boil. A more modern term for those in the former category is that of "steam generator". See **Cornish Boiler; Economic Boiler; Horizontal Boiler; La Mont Boiler; Lancashire Boiler; Locomotive Boiler; Magazine Boiler; Marine Boiler; Package Boiler; Sectional Boiler; Supercritical Once-through Boiler; Vertical Boiler; Waste-heat Boiler; Water-tube Boiler; W.I.F. Boiler.**

Boiler and Turbine Unit Plant. A boiler and turbine having no interconnections with adjacent boilers and turbines; self-contained.

Boiler Availability. The number of days per year that a boiler remains in service without shutdown for cleaning or overhaul, expressed as a percentage.

Boiler Cleaning Methods. Methods for the removal of internal scale in boilers and external deposits which impede heat transfer and may lead to dangerous overheating of the metal. The tendency to scale formation is reduced by proper boiler water conditioning. Sometimes substances may simply be added to the water causing the small amount of residual hardness to precipitate in a flocculent form, which allows it to be removed by blow-down. In many instances, however, external softening is necessary before the treated water goes forward to the boiler. Conditioning may be by a *lime* *soda process*, or an *ion exchange process*, qq.v. Scale is generally so hard as to resist a wire brush and chipping hammers, and specially shaped chisels and scrapers must be used. For water tube boilers, rotary chippers are used in which hammers are thrown against the tube. Inhibited acid cleaners may be used to assist with the removal of scale. Sometimes compressed air guns may be employed which blast out air in short bursts at high pressure. These bursts cause shock waves which break deposits from heating surfaces. External cleaning consists chiefly in the removal of soot from heating surfaces. See **Soot Blower; Soot Blowing; Water Lancing.**

Boiler Drum. A vessel usually solid forged or fabricated by fusion welding which provides a reserve of water for the tubes of a *water-tube boiler*, q.v., and as a collecting space for the steam that is generated.

Boiler Efficiency. The amount of heat in useful form produced by a boiler expressed as a percentage of the potentially useful heat put into it, i.e.:

$$\text{Boiler efficiency } \% = \frac{\text{Heat output}}{\text{Heat Input}} \times 100$$

where; heat input = weight of fuel × calorific value per unit weight;

heat output = weight of steam × (total heat/lb steam − heat/lb feed-water)

For hot water boilers;

heat output = weight of water leaving boiler × (outlet temperature − inlet temperature).

Boiler Feed Pumps and Injectors. Devices for feeding water into a boiler which is under pressure: the main methods are the use of the (a) *injector*, (b) *displacement pump*, (c) *centrifugal pump*, qq.v. The *injector*, q.v., is suitable for the smaller boiler only.

Boiling Water Reactor. A *nuclear reactor*, q.v., using water as the *moderator* and *coolant*, qq.v. The water boils under pressure in the reactor and the steam may be used to drive a *steam turbine*, q.v.; the steam being radioactive the turbine must be shielded. See **Biological Shield.**

Bomb Calorimeter. Apparatus used to determine accurately the gross calorific value of oil or coal. One gramme of oil, or finely powdered coal, is placed in a small platinum or silica crucible and inserted in a strong stainless steel "calorimetric bomb" with an air-tight cover. The bomb is filled with oxygen to a pressure of about 25 atmospheres and immersed in a calorimeter vessel containing water. The oil or coal is ignited by passing a momentary current of electricity through a thin wire within the bomb. The heat evolved by the combustion of the oil or coal is measured by the rise in temperature produced, proper corrections being applied for the water equivalent of the bomb and calorimeter vessel and for temperature changes due to radiation. The gross calorific value is then calculated.

Bonded Deposits. Deposits formed on the heating surfaces of furnaces and auxiliary plant consisting of the bonding of *fly ash*, q.v., with fusible materials produced during *combustion*, q.v. See **High Temperature Deposits; Low Temperature Deposits.**

Bone and Wheeler Apparatus. Apparatus used in the analysis of fuel gases.

Bone Seeker. A radioactive element which tends to be deposited preferentially in the bones of a body; e.g. strontium, radium or plutonium.

Boron. B. A non-metallic *element*, q.v., at. no. 5, at. wt. 10·82. It has a high neutron-capture cross-section, and boron steel is often used for control rods in nuclear reactors.

Bottled Gas. See **Liquefied Petroleum Gases** (L.P.G.).

Bottom Sediment and Water (B.S.W.). Description of contaminants of fuel oil; the greatest amounts of sediment and water are found in residual fuel oils.

Bottoms. Liquid which collects in the bottom of a vessel, during a distillation process or while in storage; hence "tower bottoms", "tank bottoms".

Boundary Lubrication. A state of partial lubrication existing between two surfaces in relative motion in the absence of full film or fluid lubrication, due to the existence of an adsorbed monomolecular layer of lubricant on the surfaces; thus a static load may squeeze out all the lubricants except that held by adsorption on the metal surfaces. See **Lubrication.**

Bourdon Gauge. A *pressure gauge*, q.v., commonly fitted to steam boilers. It comprises a tube, one end of which is open for connection to the boiler, the other end being sealed. The tube is coiled and the sealed end connected to the pointer mechanism of the gauge. When steam is admitted, the pressure inside the tube tends to straighten it; the effect of this is a movement of the sealed end and pointer mechanism. The gauge is connected to the boiler by means of a siphon containing condensate which prevents live steam from coming into contact with the gauge.

Boyle's Law. A gas law which states that the volume of a given mass of a perfect gas varies inversely as its absolute pressure, providing that the temperature remains constant. Thus, at constant temperature:

$$P_1 V_1 = P_2 V_2$$

where P_1 and P_2 represent two different pressures, and V_1 and V_2 the volumes corresponding to those pressures. See **Charles's Law; Gas Laws; Perfect Gas.**

Boys Calorimeter. A *gas calorimeter*, q.v., in use in Britain in which the *calorific value*, q.v., of a known volume of gas is calculated from the increase in temperature of a measured volume of water heated by the gas under test. The hot gases from the burner in the instrument pass over nine turns of copper tubing; the cooled gases leave the instrument through a number of holes in the lid. The temperature of the water is measured at the inlet to, and outlet from, the instrument. Both the gross and net calorific value can be calculated.

Bradford Breaker. A machine for breaking coal to a predetermined size while rejecting large refuse and tramp iron. It comprises a large cylinder rotating at about 20 rev/min; the cylinder consists of steel screen plates, the size of the screen openings determining the size of the crushed coal. The coal is broken by dropping; as it enters at one end of the cylinder it is picked up by lifting shelves and carried up until the angle of the shelf permits the coal to drop. See **Pick Breaker.**

Brass. An alloy of copper and zinc, though other elements such as aluminium, iron, lead, manganese, nickel and tin are often added; there are many varieties.

Brasses. Mineral impurities in coal which have a yellow metallic appearance; the impurities consist mainly of iron sulphides.

Breathing. The movement of oil vapours or air in and out of the relief valves or vent lines of storage tanks due to the alternate heating and cooling of the contents.

Breeches-flued Boiler. See **Galloway Boiler.**

Breeching. The connection between a furnace or incinerator and its stack.

Breeder Reactor. A nuclear reactor in which fissile material is produced by *breeding*, q.v.

Breeding. The process of producing fissile material from a fertile material such as uranium 238 or thorium 232.

Brick Kiln. See **Belgian Grate-fired Kiln; Hoffmann Kiln; Intermittent Kiln.**

Bridge-wall. A partition wall between chambers or sections of a flue over which pass the products of combustion.

Bright. Applied to lubricating oils, a term meaning clear or free from moisture. In " blowing bright " air is used to carry off moisture.

British System. A system of measurement in general use by engineers. Length is measured in inches, feet and yards; weight in ounces, pounds and tons; and time in seconds, minutes and hours. See **International System of Units (SI); Metric System; M.K.S. System.**

British Thermal Unit (Btu). The amount of heat required to raise the temperature of one pound of water through one degree Fahrenheit, from 60 to 61° F. The *mean* British Thermal Unit is $\frac{1}{180}$ of the heat required to raise the temperature of 1 lb of water from 32° to 212° F (without conversion of vapour). The following relationships apply:

$$1 \text{ Btu} = 251 \cdot 996 \text{ cal}$$
$$1 \cdot 8 \text{ Btu} = 1 \text{ Chu}$$
$$1 \cdot 8 \text{ Btu} = 453 \cdot 59 \text{ cal}$$
$$100{,}000 \text{ Btu} = 1 \text{ therm}$$
$$1 \text{ Btu} = 1055 \cdot 06 \text{ joule}$$
$$1 \text{ Btu} = 778 \cdot 169 \text{ ft lbf}$$

See **Centrigrade (Celsius) Heat Unit; Joule.**

Brown Coal. *Coal*, q.v., representing an early stage in the " coalification " of *peat*, q.v.; it is brown to black in colour, devoid of lustre, contains appreciable quantities of *volatile matter*, q.v., and has a moisture content ranging from 30 to 50 per cent. Textures vary in appearance from that of fibrous and woody to that of true coal.

Brown Smoke. Smoke produced by tarry *volatile matter*, q.v., given off at relatively low temperatures. Unless burnt it appears at the chimney top as brown smoke. See **Black Smoke.**

B.S.W. See **Bottom Sediment and Water.**

Btu Meter. A device for measuring the heat consumption between the flow and return pipes of a heating system. Varying in design a typical example comprises the following parts: (a) a flowmeter transmitter connected to an *orifice plate*, or *venturi tube*, qq.v.; (b) two electrical resistance thermometers in pockets and in common with the flowmeter transmitter connected to a Btu integrator; (c) a Btu integrator incorporating an electrical bridge network and a motor-driven chopper-bar integrating mechanism with reading scale and counting dial; (d) a flow indicator measuring water flow.

Bulk Density. The weight of a fuel per cubic foot. Examples, in lb/ft^3 are: anthracite, 47 to 51; bituminous coal, 44 to 50; hard coke, 28 to 30; gas coke, 22 to 26; fuel oil, 49 to 61.

Bulk Supply Tariff. A preferential *tariff*, q.v., for the purchase of supplies of a commodity or service in bulk. The *Central Electricity Generating Board*, q.v., sells electricity in bulk to the Area Electricity Boards who then resell it to the consumer. The tariff is based upon the aggregate costs of generation incurred by the Generating Board, and the surplus or " balance of revenue " to be earned. It is a *two-part tariff*, q.v., with a kilowatt demand charge and a running charge per kilowatt hour.

Bunker " C " Fuel Oil. A heavy residual fuel oil supplied to ships and industry.

Bunker Coal. Coal supplied as fuel to ships.

Bunsen Burner. A *gas burner*, q.v., in which air is inspirated by the gas and mixed in the bunsen tube; the fuel and oxygen are heated together, the result being a short blue flame with little or no luminescence. The hydrocarbons and oxygen form hydroxylated compounds before ignition. If combustion is incomplete aldehydes can be detected by their faint acrid odour. See **Fishtail Burner.**

Burn Up. The amount of the fissile material in a *nuclear reactor*, q.v., which is destroyed by fission or neutron capture, expressed as a percentage of the original quantity of fissile material present; or alternatively, the heat obtained per unit mass of fuel, expressed generally as MW/days tonne. In nuclear reactors using metallic uranium fuel elements an average burn up of 3000/4000 MW/days tonne is obtained; with uranium dioxide elements in advanced gas-cooled reactors, an average burn-up of 18,000 MW/days tonne is expected.

Burner. A device which produces a flame. A burner must mix the fuel and oxidizing agent in proportions suitable for *ignition*, q.v., and steady combustion; and it must supply the mixture at rates which ensure the stability of the flame. See **Gas Burner; Oil Burner.**

Burning Area. The total area of a grate or hearth, or combination thereof on which combustion takes place.

Burning Oil. Paraffin oil or *kerosine*, q.v.

Burning Rate. The amount of fuel or waste consumed, usually expressed as pounds per square foot of *burning area*, or *grate*, q.v.

Burst Can (Cartridge or Slug). A fuel element can in which a leak has developed, permitting the escape of radioactive fission products.

Butane. C_4H_{10}. A paraffin hydrocarbon with a b.p. of 32° F (0° C); sp. gr. of liquid at 60° F of 0·57; a gross calorific value of 3200 Btu/ft^3 dry and 21,300 Btu/lb (net 3000 Btu/ft^3 dry and 19,650 Btu/lb). The sulphur content is negligible with a maximum of 0·01 per cent by weight. There are 477 therms/ton (gross c.v.) and 31 ft^3 air is required to burn 1 ft^3 of gas. Butane burns in air to carbon dioxide and water. Butane (perhaps with an admixture of 7 per cent propane) is available in cylinders—hence the term "bottled gas"—suitable for domestic and commercial use. It may be used also for the enrichment of *town gas*, q.v.; heating processes requiring special atmospheres; and the fuelling of fork-lift trucks. In normal atmospheric conditions butane requires pre-heating slightly, while *propane*, q.v., will gasify immediately. See **Liquefied Petroleum Gases.**

Butterfly Damper. An adjustable, pivoted, plate normally installed in the flue between furnace and stack. See **Damper.**

By-product Fuels. Fuels produced incidentally in the manufacture of a product. Examples are bark, black liquor, bagasse, sawdust, blast furnace and coke oven gas; plutonium is a by-product of nuclear power generation.

C

Cadmium. Cd. A white metallic *element*, q.v., at. no. 48, at. wt. 112·40. The metal is used as a constituent of fusible alloys and as an anti-corrosive coating for metal articles. It has a high-capture cross-section for neutrons with energies less than 0·5 electron-volt; it has been used occasionally as a neutron absorber in control rods and special shields in nuclear reactors.

Caesium. Cs. An alkali metal, at. no. 55, at. wt. 132·91. The isotope Cs^{137} is a fission product of a *nuclear reactor*, q.v., fuelled with uranium; it is a radioactive gamma-ray emitter with a *half-life*, q.v., of thirty years.

Caking Coal. A coal which liberates, at temperatures of the order of 645 to 840° F (340 to 450° C), substances which cause the main mass of the coal to become more or less fluid or "plastic". As the coal becomes plastic, gases are evolved within the material which set up a pressure; as the gases escape the material becomes cellular. At higher temperatures of 840 to 930° F (450 to 500° C) decomposition proceeds rapidly and the coal hardens into coke. See **Non-caking Coal**.

Calcium Silicate. An insulating material, but not suitable for surfaces above 1400° F (760° C); it is light in weight, moisture resistant and it possesses a low thermal conductivity.

Calorie. The amount of heat required to raise the temperature of one gramme of water through one degree Centigrade, i.e. from 14·5 to 15·5° C. One calorie = 4·1868 *joule*, q.v.

Calorific Value. The quantity of heat developed by the complete combustion, q.v., of a given weight or volume of fuel; sometimes referred to as "heating value". It is expressed in terms of British thermal units per pound (Btu/lb), Centigrade (Celsius) heat units per pound (Chu/lb), calories per gramme (cal/g), or as kilogramme calories per kilogramme (kcal/kg) for solid fuel; and similarly for liquid fuel or cubic feet of gaseous fuels. A good house coal has a calorific value of about 13,000/14,000 Btu/lb; coke about 12,000 Btu/lb; fuel oil about 18,000/19,000 Btu/lb; town gas about 450 to 550 Btu/ft³. A kilowatt hour of electricity generates 3412 Btu. The full heat value is described as the higher or gross calorific value. In practice steam released in the process of combustion (from inherent moisture and hydrogen in the fuel) is not condensed and *latent heat*, q.v., is lost; the effective heating value is therefore lower, this being known as the lower or net calorific value. See **Gross Calorific Value; Net Calorific Value.**

Calorifier. A vessel for supplying hot water; it is essentially a heat exchanger, the source of heat being a steam or hot water pipe coil running through the vessel.

Calorimeter. Any vessel used for containing a liquid during heat experiments; it is generally essential to calculate its *water equivalent*, q.v.

Can. A container in which the fuel rods of a *nuclear reactor*, q.v., are sealed to: (a) prevent the escape of fission products into the *coolant*, q.v.; (b) to protect the fuel rods from corrosion by the coolant.

Candela. Unit of luminous intensity; the magnitude of the candela is such that the luminance of the total radiator, at the temperature of solidification of platinum, is 60 candelas per square centimetre (60 cd/cm²).

Cannel Coal. A coal derived from *sapropel*, q.v., typically dark-brown to black in colour with a dull, greasy lustre. Cannel is composed predominantly of minute *vitrinite* particles, with little or no admixture of *micrinite* or *fusinite*, qq.v. It is highly volatile, up to 56 per cent, and usually non-caking. The calorific value may be as high as 15,700 Btu/lb on a dry ash-free basis. See **Boghead Coal.**

Capacitor. An arrangement of conductors in the form of either sheets or foil separated by a thin dielectric. It provides what is known as "capacity", and is capable of storing a comparatively small electric charge.

Capacity. The estimated maximum level of production from a plant on a sustained basis, allowing for all necessary shut-down, holidays, etc. The capacity factor is calculated:

$$\text{Capacity factor} = \frac{\text{Output for period}}{\text{Rated capacity} \times \text{hours in period}}$$

See **Availability; Installed Capacity; Load Factor, Plant; Refinery Capacity.**

Capture. The process in which an atom or nucleus acquires an additional particle; for example, the capture of neutrons by nuclei which often results in the production of a radioactive isotope of the capturing element.

Car Tunnel Kiln. A continuous ceramic kiln comprising a tunnel some 100 to 300 ft in length, a furnace being situated at about the centre of its length. Rail tracks are laid through the kiln and on these special cars carry the ware to be fired. The ware passes slowly through preheating, firing and cooling zones. Car tunnel kilns are fired by town gas, electricity, fuel oil or producer gas; the conversion from coal-fired intermittent kilns to continuous car tunnel kilns has been a major factor in increasing productivity and reducing smoke emission in the pottery industry.

Carbon. C. A non-metallic *element*, q.v., at. no. 6, whose allotropic modifications include diamond, graphite, charcoal and coke. When a sufficient amount of oxygen is available, carbon may be burned completely to *carbon dioxide*, q.v., the reaction being expressed $C + O_2 \rightarrow CO_2$. The calorific value of carbon is 14,450 Btu (8028 Chu)/lb. Its low neutron-capture cross-section and low atomic mass render it suitable, in the form of graphite, for use as a moderator in a thermal nuclear reactor to slow down the fast neutrons produced in fission. It is also used as a neutron reflector.

Carbon Black. Fine, fluffy carbon particles resulting from the incomplete combustion of either gaseous or liquid hydrocarbons. It is used as a reinforcing agent in rubber and as a pigment and filler in plastics, paint, varnish, ink and paper. The particles are essentially pure carbon ranging in size from $2 \cdot 5 \times 10^{-6}$ to 350×10^{-3} in. diameter. Carbon black is produced in a reducing atmosphere. In the oil furnace process either a part of the feedstock or an auxiliary fuel is burned in the furnace to generate the heat required to decompose the oil into carbon black and gas.

Carbon Dioxide. CO_2. A colourless gas produced when carbon is burned in a sufficient supply of oxygen to complete the reaction $C + O_2 \rightarrow CO_2$. It is present in flue gases to a varying extent, but in typical cases amounts to 12 per cent of the gases by volume. If the supply of combustion air is insufficient, *carbon monoxide*, q.v., is formed. Carbon dioxide is a normal constituent of the atmosphere to the extent of about $0 \cdot 03$ per cent by volume.

Carbon Dioxide Recorder. An instrument of the thermal conductivity type. *Carbon dioxide*, q.v., and water vapour possess similar thermal conductivities which are about half those of oxygen and nitrogen; if, therefore, a sample of flue gas is passed over one arm of a heated Wheatstone Bridge circuit, the "out-of-balance" set up in the circuit is a measure of the CO_2 percentage in the flue gases. See **"Fyrite" CO_2 Apparatus; Orsat Apparatus.**

Carbon Disulphide. CS_2. A colourless liquid, when pure, turning yellow on exposure to air; it is extremely inflammable. The unpleasant characteristic smell of commercial carbon disulphide is due to its impurities. It is used extensively in many industries as a solvent for sulphur and rubber. See **Carpenter-Evans Process.**

Carbon Monoxide. CO. An invisible, tasteless, odourless and highly poisonous gas. It burns with a pale blue flame to form *carbon dioxide*, q.v. It results from the incomplete combustion of carbonaceous materials, being encountered in many industrial works and occurring in the exhaust gases from internal combustion engines and in smoke from chimneys. Carbon monoxide has an affinity for the haemoglobin of the blood some 300 times that of oxygen and readily forms carboxyhaemoglobin in the red blood corpuscles. As the red blood corpuscles cannot carry their full quota of oxygen to the tissues of the body, the tissues suffer from oxygen starvation and show the symptoms of "carbon monoxide poisoning". Significant concentrations of carbon monoxide, expressed in parts per million by volume, are: maximum allowable concentration in an atmosphere in which a man works for 8 hours at a time, 100;

headache after 3 to 4 hours, 200; severe headache, palpitation, nausea and dizziness, after 1 to $1\frac{1}{2}$ hours, 500; prolonged exposure may be fatal, 1000. See **Combustion.**

Carbon Residue. The quantity of solid deposits obtained when medium and heavy fuel oils are subjected to evaporation and *pyrolysis*, q.v., at elevated temperatures. While the bulk of the sample evaporates, the residue decomposes and forms carbonaceous deposits. This property is of importance for oils used in burners, gas-making, compression ignition engines and certain lubricating oils. See **Conradson Test; Ramsbottom Test.**

Carboniferous Period. A period, between 200 and 250 million years ago, in which conditions suitable for the accumulation and preservation of coal-forming deposits were most widely established over the earth's surface.

Carbonization. The *destructive distillation*, q.v., of coal to produce coke, gas and liquid products. Low temperature carbonization means carbonization at temperatures ranging up to 1470° F (800° C); high temperature carbonization means carbonization at temperatures above 1470° F (800° C). Coking coals with more than 33 per cent *volatile matter*, are used for the manufacture of *town gas* and *coke*, qq.v.; coal is carbonized in horizontal, intermittent vertical and continuous vertical retorts. See **Continuous Vertical Retort; Horizontal Retort; Intermittent Vertical Retort.**

Carbonized Briquettes. Briquetted solid fuels *carbonized*, q.v., in the process of manufacture. See "**Phurnacite**".

Carboxylic Resin. A cation exchange resin; it has the property of accepting calcium and magnesium ions in exchange for hydrogen. Used as an additional stage in water softening if water is too alkaline for base exchange softening; it removes the bicarbonates. The hydrogen combines with the carbonate ions to produce carbonic acid. Water leaving this de-alkalization unit contains virtually no alkaline hardness. The carbonic acid is subsequently removed.

Carburetted Water Gas. Water gas enriched with gasified hydrocarbon oil; about two gallons of gas oil may be used per 1000 ft^3 of gas. The calorific value is about 500 Btu/ft^3, similar to that of *town gas*, q.v. The gas burns with a yellowish luminous flame. This type of plant has often been used to meet winter peak loads. A typical analysis, the constituents being expressed in percentages, is: hydrogen, 36; carbon monoxide, 30; carbon dioxide, 6; nitrogen, 5; methane, 15; hydrocarbons, 8.

Carburetting. In *town gas*, q.v., manufacture, the enrichment of

38

gas by the addition of gaseous hydrocarbons, e.g. from gas oil cracked in a carburettor.

Carburettor. A device for producing a suitable fuel/air mixture for a *gasoline engine*, q.v., under a variety of operating conditions. Small quantities of fuel are drawn from a jet by a current of air passing through a venturi in which the jet is situated; the gasoline is then atomized and partly vaporized. By the use of a *choke*, q.v., and compensating jets the fuel composition can be varied to meet engine requirements. Also a device for cracking *gas oil*, q.v.

Carburizing. A case hardening or cementation *heat treatment*, q.v., process for raising the carbon content of steel by heating in a carbonaceous medium. *Carbon monoxide*, q.v., carburizes and is essential for the heat treatment of high carbon steels. See **Decarburization.**

Carcinogenic Compounds. Complex chemical compounds producing cancer in experimental animals and strongly suspected of contributing to lung cancer in man. One of the best known carcinogens is 3,4 benzpyrene. In 1933, Cook and co-workers, with the assistance of fluorescence spectroscopy, isolated from two tons of pitch a substance characterized by synthesis as 3,4 benzpyrene; it is now known that coal tar contains about 1·5 per cent of the carcinogen. In recent years, *benzpyrene*, q.v., and other carcinogenic hydrocarbons have been identified in soot, carbon black, processed rubber, the exhaust gases of gasoline and diesel engines, coal gas, coffee soots, smoked food, and cigarette and tobacco smoke. Relatively large concentrations of some carcinogenic hydrocarbons have also been found in the atmospheres of some cities. Some of the complex polycyclic hydrocarbons formed in smoke have been shown to be carcinogenic when applied to the skin of animals. Coal tar and certain petroleums become carcinogenic only after being heated.

Carnot Cycle. An ideal cycle for a *heat engine*, q.v., giving maximum theoretical efficiency; no practical cycle is able to equal the Carnot cycle. In the ideal cycle, heat is added at a constant temperature T_1 and rejected at a lower temperature T_2:

$$\text{Thermal efficiency} = \frac{\text{Work done}}{\text{Heat added}}$$

$$= \frac{\text{Heat added} - \text{Heat rejected}}{\text{Heat added}}$$

$$= \frac{T_1 - T_2}{T_1}$$

Carpenter-Evans Process. A process for the removal of *carbon disulphide*, q.v., from coal gas or petroleum oil gases; the gas stream passes over a nickel catalyst at about 840° F (450° C), carbon disulphide being converted to *hydrogen sulphide*, q.v. The hydrogen sulphide is then removed in a separate process. See **Hydrogen Sulphide Removal.**

Carpet Losses. Losses which occur when solid fuel is stored outside directly on the earth; these can amount to 3 per cent of the initial amount stored.

Carry-over. Water droplets and impurities carried by steam from a boiler to the superheater.

Cartridge. The fuel element of a *nuclear reactor*, q.v., consisting of a rod of fuel, such as uranium, hermetically sealed in a container or can. See **Burst Can.**

Cascade Heat Exchanger. A *heat exchanger*, q.v., which uses small refractory particles as a heat carrier. The particles are heated by a counter-current flow of hot gases in one chamber, and then pass to a second chamber where they give up their heat to air or any other gas that is to be preheated.

Casinghead Gasoline. *Gasoline*, extracted from *natural gas*; also known as *natural gasoline*, q.v.

Cast-iron. An alloy of iron and carbon; a typical analysis of rainwater goods, the constituents being expressed in percentages, is: carbon, 3·4; silicon, 2·5; manganese, 0·7; phosphorous, 1·0; sulphur, 0·12; balance, iron. See **Grey Cast-iron; Malleable Cast-iron; White Cast-iron.**

Catalyst. A substance which accelerates or retards the rate of a chemical reaction, without itself undergoing any permanent change in its composition and which is normally recoverable when the reaction is completed.

Catalytic Cracking. See Fluid Catalytic Cracking Unit (FCCU).

Catalytic Rich Gas (CRG) Process. A *Gas Council*, q.v., pressure process for manufacturing rich gas from *naphtha*, q.v.; it has a calorific value of 650 to 800 Btu/ft³. The gas may be supplied direct to the consumer after blending.

Caustic Embrittlement. A cause of metal failure occuring in steam boilers at riveted joints and at tube ends, believed to result from the action of certain constituents of concentrated boiler water upon steel under stress.

Cell. The unit of a battery. A "primary" cell consists of two dissimilar elements (usually carbon and zinc), known as "electrodes", immersed in a suitable liquid or paste known as the "electrolyte". Such a cell will produce a direct current of 1 to 1·5 volts. A "secondary" cell or accumulator is of somewhat similar design, but is made suitable for use by passing a direct current of correct strength through it in a certain direction. When "charged" in this way, current may be obtained from it for a number of hours. Each cell of a lead-acid accumulator produces approximately 2 volts; a 12 volt car battery contains 6 cells. See **Fuel Cell.**

Celsius. A temperature scale with two fixed points, 0 being the freezing point of water and 100 the boiling point of water at normal atmospheric pressure. Formerly called Centigrade, the use of that name for this scale was abandoned by the General Conference on Weights and Measures in 1948. The abbreviation ° C remained the same. However, the term Centigrade remains in fairly common use. See **Centigrade (Celsius) Heat Unit; International Temperature Scale.**

Celsius Heat Unit. See **Celsius; Centigrade (Celsius) Heat Unit; International Temperature Scale.**

Cement Kiln. A rotating steel cylinder up to 600 ft in length and to 10 ft in diameter, lined with alumina bricks. The cylinder is slightly inclined to the horizontal so that the contents may gradually descend as the cylinder rotates. The kiln is heated by pulverized coal or oil, temperatures of the order of 2370° F (1300° C) being attained. Chalk or limestone, and clay or shale, mixed with water in the required proportions, are fired in the kiln to form cement clinker. The clinker is ground, with the addition of about 5 per cent gypsum, to form cement. The slurry of lime and clay enters the upper end of the kiln, passing through three zones. In the first zone it is dried; in the second zone the chalk loses its *carbon dioxide*, q.v., and becomes quicklime; in the third or lower zone the oxides of calcium, alumina and silicon combine to form cement clinker.

Centigrade (Celsius) Heat Unit (Chu). The amount of heat necessary to raise the temperature of one pound of water through one degree Centigrade, or more exactly $\frac{1}{100}$ part of the amount of heat required to raise the temperature of one pound of water from 0 to 100° C. The following relationships apply:

$$1 \text{ Chu} = 1·8 \text{ Btu}$$
$$= 453·592 \text{ cal}$$
$$= 1899·10 \text{ joule}$$
$$= 1400·71 \text{ ft lbf}$$

See **British Thermal Unit; Calorie; Celsius; Joule.**

41

Centigrade (Celsius) Scale. A temperature scale based on two fixed points: (a) the lowest (0) being the temperature of melting ice and (b) the highest (100) being the boiling point of water at normal atmospheric pressure. The Centigrade (Celsius) and Fahrenheit scales are conventional scales:

$$0° \text{ C} = 32° \text{ F}; \ 100° \text{ C} = 212° \text{ F},$$

where 1 Centigrade (Celsius) degree = 1·8 Fahrenheit degrees. See **Temperature Scales.**

Centistokes. An international unit for the measurement and expression of all kinematic viscosity values. The equivalents are:

$$100 \text{ centistokes} = \begin{cases} 1 & \text{stokes} \\ 407 \text{ sec} & \text{Redwood No. 1} \\ 460 \text{ sec} & \text{Saybolt Universal} \\ 46 \text{ sec} & \text{Saybolt Furol} \\ 13·3 & \text{Degrees Engler.} \end{cases}$$

Central Electricity Generating Board (C.E.G.B.). A central body established under the Electricity Act of 1957, the Board's statutory responsibilities are "to develop and maintain an efficient, co-ordinated and economical system of supply of electricity in bulk for all parts of England and Wales". The Board owns and operates some 230/240 power stations, both conventional and nuclear, and the main transmission lines for the bulk supply of electricity to twelve Area Electricity Boards.

Centrifugal Fan. Or radial flow fan. A *fan*, q.v., in which air flows into the "eye" and is discharged from the periphery at right angles to the direction of entry. See Fig. 3.

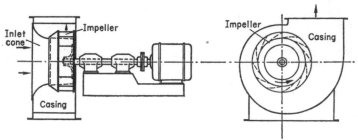

FIG. 3. *A centrifugal fan.*

Centrifugal Pump. A pump using the principle of centrifugal force; the liquid enters the "eye" or centre of a rotating impeller through which it flows radially under centrifugal force leaving through an exit at the periphery; in this way the kinetic energy of the liquid is converted into pressure energy. This type of pump may be used as a boiler feed pump; it is most suitable for higher pressures and outputs. These pumps are usually electrically driven; some of the larger sizes are turbine driven, the exhaust steam being utilized for heating the feed-water.

Cerenkov Radiation. Visible radiation emitted when charged particles move through a transparent medium with a velocity greater than the velocity of light in that medium; Cerenkov radiation is observable as a faint blue glow in the vicinity of the fuel elements of a *swimming pool reactor*, q.v.

Cetane Number. A rating for diesel fuels comparable to the octane rating for gasolines. It is the percentage of cetane, $C_{16}H_{34}$, which must be mixed with alpha-methyl naphthalene to give the same ignition performance, under test conditions, as the fuel under examination. Thus it is a measure of the time required for a liquid fuel to ignite after injection into a compression ignition engine. On an empirical scale 0 to 100, a short delay period is indicated by a cetane number from 40 to 70; a long delay by a cetane number below 40. Most diesel engines operate satisfactorily on fuels in the 40 to 50 cetane number range. See **Octane Number.**

CFR Engines. Cooperative Fuel Research engines used to measure the anti-knock quality of production and experimental fuels, including motor, aviation and automotive diesel fuels; engines developed by the Coordinating Fuel Research Council in America and used by all the Oil Companies for this purpose.

C.G.S. System. See **Metric System.**

Chain Grate. An endless chain of short bars linked together to carry a fuel bed through a furnace.

Chain Grate Stoker. A *mechanical stoker*, q.v., consisting of an endless chain of cast-iron links driven through sprockets and moving from front to rear; the fuel is fed by gravity, being spread over the grate to a thickness regulated by the fire door. The fuel burns as the grate moves along. At the rear end, ash plates remove the ashes and clinker from the grate. The first 3 ft or so of the grate is shrouded by a refractory ignition arch which extends down to the level of the *chain grate*, q.v., on either side. The arch is heated by the burning fuel beneath it and so acts as a source of radiation to stabilize ignition. The fire door or guillotine is adjustable so that the thickness of the fuel bed may be varied to suit the conditions, e.g. the larger the fuel the thicker should be the fuel bed. The fuel bed is normally about 4 in. thick. The stoker may be operated with natural, induced or forced draught. In the large units combustion rates of 40 lb/ft² are attainable.

Chain Reaction. A self-sustaining process; in a *nuclear reactor*, q.v., the neutron-induced fission of uranium 235 results in the liberation of neutrons which cause the fission of further uranium 235 atoms.

Char. The by-product solid fuel from low temperature *carbonization*, q.v., processes; it is the residue after extracting from coal valuable tars, bitumen and gases for use as raw materials in the chemical industry.

Characterization Factor, K. For crudes and petroleum fractions, the ratio of the cube root of the normal boiling point in degrees Rankine to the specific gravity at 60° F (15·6° C). United States crudes appear to lie in the range 11·5 to 12·5.

Charcoal. The solid residue, black in colour, of the *destructive distillation*, q.v., of wood; it is a brittle porous material retaining the original shape of the wood while its micro-structure preserves the vegetable cell structure. It burns without smoke and is used for heating and cooking; in the United States, and elsewhere, it is used for grilling and barbecuing. It is also used in the manufacture of gunpowder, as an absorbent and decolourizing agent, in hop drying, and in work where a very pure fuel is required. Activated charcoal (produced by blowing steam through charcoal) adsorbs organic compounds; it is used in sugar refining and solvent recovery.

Charles's Law. A gas law stating that the volume of a given mass of a perfect gas varies directly as the absolute temperature, providing the pressure remains constant. Thus, at constant pressure:

$$\frac{V_1}{T_1} = \frac{V_2}{T_2}$$

Where T_1 and T_2 represents the two different temperatures, and V_1 and V_2 the volumes corresponding to those temperatures. See **Boyle's Law; Gas Laws; Perfect Gas.**

Chelating Agents. Substances suitable for removing the residual hardness from soft or softened waters; they take polyvalent metal ions, such as calcium and magnesium, into their molecular structure and prevent them from precipitating; the effect is the same as if the hardness had been removed and there is consequently no precipitation or scale formation.

Chemical Energy. The energy liberated in a chemical reaction, as in the combustion of solid, liquid and gaseous fuels.

Chequer-Brickwork. Multiple openings in a baffle wall to permit combustion gases to pass yet promote turbulent mixing of the products of combustion. In a *regenerator*, q.v., serving a steel or glass furnace chequer-brickwork is utilized as a means of storing heat.

Chimney Heights, Memorandum on. A Memorandum issued by the Ministry of Housing and Local Government and the Scottish Development Department in 1963 which provides a simple method of calculating the height of chimneys to give satisfactory ground level concentrations of sulphur dioxide. The method takes account of the maximum rate of emission of sulphur dioxide as calculated from the sulphur content of the fuel and the maximum rate of burning, the character of the surrounding area and the height of adjacent buildings. The scope of the memorandum is limited to chimneys serving industrial boiler plant of maximum continuous rating greater than approximately 650 lb steam/h (or 0·65 million Btu/h) and not exceeding 450,000 lb steam/h (or 450 million Btu/h); and to the chimneys of furnaces burning fuel at a maximum rate greater than approximately 100 lb/h of coal or 50 lb/h of oil and not exceeding 50,000 lb/h of coal or 30,000 lb/h of oil. The first stage of calculation gives the "uncorrected chimney height". The next stage of calculation is the amendment of this uncorrected height to allow for the height of surrounding buildings and so obviate the risk of *down-draught*, q.v. Appendices I to VI to the Memorandum contain nomograms to enable both stages of calculation to be carried out without difficulty.

Chlorination. The addition of chlorine in gaseous form to power plant circulating water to control slime formation in condensers, and mussel growth in culverts.

Chlorine. Cl. An *element*, q.v., in coal, at. no. 17, at. wt. 35·453; it occurs in varying amounts, chiefly as sodium chloride. The average amount present in British coals is about 0·23 to 0·24 per cent, with an upper limit of 0·75 per cent. In all combustion equipment, at least 90 per cent of the chlorine in the coal appears to be emitted as hydrochloric acid in the flue gases; in certain conditions this acid can give rise to serious corrosion problems and high temperature bonded deposits on heating surfaces, particularly in *water-tube* *boilers*, q.v. A heavy residual fuel oil may contain about 0·004 per cent of chlorine.

Choke. A device for enriching a fuel/air mixture to enable a cold engine to start.

Chromatography. A technique for separating and analyzing mixtures of chemical substances; a flow of solvent or gas causes the components of a mixture to migrate differentially from a narrow starting zone through a special medium. Separations are usually made in columns of selected sorptive powders or liquids, or in strips of fibrous media such as paper. In gas-liquid partition chromatography (GLC), the column packing is a liquid solvent distributed on an inert solid support; in gas-solid chromatography (GSC) the column packing is a surface active sorbent such as charcoal, silica gel or activated alumina. Gas chromatography is used principally as an analytical technique for the determination of volatile compounds. The composition of the emerging gas is monitored by a suitable detecting device capable of indicating the presence and amount of the individual components; the most popular detectors are the thermal conductivity detector (katharometer) and the flame-ionization detector. See **Spectrophotometry.**

Chrome Brick. A *refractory*, q.v., containing 35 to 45 per cent of chromic oxide, 15 to 35 per cent of magnesium oxide, 10 to 25 per cent of alumina and 12 to 20 per cent of iron oxide.

Chrome-magnesite Brick. A *refractory*, q.v., containing 40 to 60 per cent of magnesium oxide and 20 to 35 per cent of chromic oxide.

Clarain. Soft humic coal which presents a shiny black, well-laminated appearance. It consists largely of *vitrinite*, and some *micrinite* with traces of *fusinite*, qq.v.

Classifier. In relation to *pulverized fuel*, q.v., a device to permit the passage of material of desired fineness while returning oversize material for further grinding.

Claus Kiln. An oil refinery unit for the recovery of sulphur from hydrogen sulphide rich gases; hydrogen sulphide is burned in an insufficient supply of oxygen for complete combustion, the principle of *preferential combustion*, q.v., giving the reaction:

$$2H_2S + O_2 \rightarrow 2H_2O + 2S$$

Clay Treating. The removal of oxygen, nitrogen and sulphur compounds from certain oils following refining; removal is effected by adsorption on attapulgus clay, Fuller's earth and other suitable materials.

Clean Air Act, 1956. See **Clean Air Legislation.**

Clean Air Legislation. National, state or city legislation introduced to control *air pollution*, q.v.; most industrial countries have now introduced legislation. Britain has two Acts of Parliament relating solely to air pollution: (a) the Alkali, Etc., Works Regulation Act, 1906; and (b) the Clean Air Act, 1956. The Alkali Act deals with emissions to atmosphere from a wide range of scheduled industrial processes including power stations, oil refineries and petro-chemical processes, the chemical industry, and the iron and steel industry. The Clean Air Act deals with emissions of smoke, grit and dust from a wide range of non-scheduled industry, and smoke from domestic dwellings; under this Act, smoke control areas are being established. The Alkali Act is enforced by a small central body of highly qualified alkali inspectors, while the Clean Air Act is implemented by the public health inspectors of local authorities. In the United States of America the primary responsibility for abatement action rests with state and local governments. The U.S. Clean Air Act, 1963 (amended in 1965), makes available federal technical and financial assistance. The Secretary of Health, Education and Welfare is authorized to: (a) encourage cooperative action by states and local government; (b) encourage the enactment of improved and uniform laws; (c) encourage agreements and compacts between states for the prevention and control of air pollution. He is required to establish a national research and development programme and to conduct investigations and surveys on specific pollution problems. Where interstate air pollution endangers human health and welfare, the Secretary is authorized to call a conference and recommend remedial action if necessary. Under the Amendment of 1965, the Secretary has authority to control air pollution from new motor vehicles; as a result all new automobiles are to be equipped with exhaust pollution controls from 1968.

"Cleanglow". A *gas coke*, q.v., of a highly reactive nature produced in continuous vertical retorts from specially selected coals; it is a smokeless fuel, very free burning, and suitable for any domestic appliance. See **Authorized Fuels; Smoke Control Area.**

Cleaning Fires. The removal of ash and clinker from a fuel bed; two methods are adopted with fixed grates: (a) the side method, in which the live coal is pushed to one side of the grate and then to the other, enabling the ash and clinker on the grate to be removed; (b) the front-and-back method in which the fire is pushed back against the bridge while the nearest parts of the grate are cleaned. As the whole grate cannot be cleaned by this method it is only

47

justified when the demand for steam is high; the side method should be employed as soon as circumstances permit.

Clinker. Hard material formed from the cooling of fused *ash*, q.v. Clinker is generally undesirable in a furnace. It obstructs the flow of primary air through the fuel bed and increases draught requirements; leads to *blow holes*, q.v., forming in other parts of the fuel bed; and its removal necessitates the use of the slice. Clinkering depends on the fusion temperature of the ash and the temperatures to which ash is exposed. Steam or water jets may be used to help reduce the formation of clinker. See **Clinker Prevention.**

Clinker Prevention. Measures adopted to prevent *clinker*, q.v., formation: these are to (a) avoid thick fires; (b) use tools sparingly; (c) avoid mixing ash with burning fuel; (d) keep the fire level by careful firing; (e) operate as long as possible without disturbing the fires and then clean out thoroughly; (f) in last resort to use water sprays under the grate.

Closed-loop Control System. Or feedback control system. A control system in which a controlled variable is measured and compared with a standard representing the desired performance. Any deviation from the standard is "fed back" into the control system so that the deviation of the controlled variable from the standard can be corrected. See **Feedback Controller; Open-loop Control System.**

Cloud Point. The temperature at which a cloud or haze appears when a sample of oil is cooled under prescribed test conditions. Clouding may be due to separated waxes or to water coming out of solution in the oil. The main use of the cloud point is to give an indication of the lowest temperature at which an oil may be used without causing the blockage of filters.

CO_2—Acceptor Process. A possible technique for the gasification of coal. Coal is gasified in steam in a fluid bed in the presence of lime; the carbon dioxide formed by the reaction between steam and carbon reacts exothermally with the lime, producing sufficient heat to support the endothermic gasification reactions; and the hydrogen-rich gas that is produced hydrogenates part of the coal substance to form methane. Lime carbonated in the process is calcined for recycling.

"CO" Boiler. A boiler which may be used as an integral part of a *fluid catalytic cracking unit*, q.v., in an oil refinery; its purpose is to provide steam for the cracker, using for this purpose hot carbon monoxide—rich waste gases from the regenerator—and such supplementary fuel as required. The burning of the carbon coating on the catalyst with air during the regenerative stage produces a waste gas containing up to 8 per cent carbon monoxide at a temperature of about 1000° F (538° C); the waste gas is discharged from the regenerator through cyclone separators to remove a high percentage of the entrained catalyst. It may then be utilized in a " CO " boiler; supplementary fuel is always necessary, however, to give a boiler furnace temperature of about 1800° F (982° C).

Coal. A general name for a firm, brittle, sedimentary, combustible rock derived from vegetable debris which has undergone a complex series of chemical and physical changes during the course of many millions of years. See **Anthracite; Bituminous Coal; Brown Coal; Lignite; Peat; Sub-bituminous Coal; Welsh Dry Steam Coal; Wood.**

Coal Classification Systems. See **A.S.T.M. Coal Classification; International Coal Classification** (E.C.E.); **Mott's Classification; National Coal Board Classification; Parr Classification; Seyler's Classification.**

Coal Cleaning. See **Baum Coal Washer; Cyclone Washer; Dense Medium Process; Dry Cleaning Process; Froth Flotation.**

Coal Equivalent. A common measure customarily used in energy statistics where the consumptions of different fuels in any economy are being compared. The " coal equivalent " of any fuel, other than coal, is the heat content of the fuel expressed as a quantity of coal with the same heat content. In British practice one ton of oil is taken as equivalent to 1·7 tons of coal, and 280 therms of natural gas as equivalent to one ton of coal, these having (on average at any rate) the same heat content. Nuclear power and hydro-electricity are equated to the amount of coal needed to produce electricity at the current efficiency of conventional power stations.

International conversion factors differ in that when oil is expressed in "coal equivalent" one ton of crude oil is taken as equal to 1·3 tons of coal and one ton of refined products as equalling 1·5 tons. For crude and refined products together, the Organisation for Economic Co-operation and Development uses a factor of 1·43 tons of coal per ton of oil.

Coal Gas. Gas obtained from the *destructive distillation*, q.v., of suitable bituminous coals in closed retorts at high temperatures.

Coal Hydrogenation. The addition of hydrogen to coal at elevated temperatures and pressures to increase the hydrogen/carbon ratio towards that prevailing in petroleum thus yielding liquid and/or gaseous hydrocarbons, notably motor spirit. Clean small coal of

Coal Hydrogenation

selected rank at low ash content is ground and mixed with recycled heavy oils to give a paste which is fed with recirculated hydrogen at about 260 atmospheres pressure to coal hydrogenation units. The paste and hydrogen are heated to about 770° F (410° C) before entering converters. Catalyst is added partly to the coal paste and partly directly to the converters. The total product of the converters is separated in a hot catchpot which collects a heavy asphaltic fraction, and a cold catchpot which collects distillate oil. After this " liquid phase " hydrogenation process some of the products pass through a " vapour phase " hydrogenation process. Petrol is washed to extract phenols, and refined. The yield of regular grade motor spirit from a suitable coal of about 84 per cent carbon and 2·5 per cent ash is 46·1 per cent, i.e., one ton of petrol requires 2·17 tons of clean coal. If the coal required for hydrogen production is taken into account a total of 4·50 tons of coal is required for one ton of petrol.

Coal Segregation. The tendency for lumps of coal to separate out from the fine coal between, say, the coal bunkers and mechanical stoker hoppers.

Coal Tar. A by-product of the manufacture of coke and coal gas; a viscous mixture consisting mainly of aromatic compounds. It has a calorific value of from 15,000 to 16,500 Btu/lb (8330 to 9165 Chu/lb). See **Coal Tar Fuels.**

Coal Tar Fuels. Fuels produced by the distillation of *coal tar*, q.v. Following distillation six standard types of coal tar fuel are produced; each grade is identified by the degree of preheat (° F) required for optimum atomization. The six grades are: C.T.F. 50, grade 'A' creosote; C.T.F. 100, grade 'B' creosote; C.T.F. 200, pitch/creosote mixture; C.T.F. 250, heavy pitch/creosote mixture; C.T.F. 300, heavy pitch/creosote mixture; C.T.F. 400, pitch. The gross calorific value of these fuels ranges from 16,500 Btu/lb (9200 Chu/lb) for C.T.F. 50 to 16,300 Btu/lb (9050 Chu/lb) for C.T.F. 400. The carbon/hydrogen ratio is high in all grades. The sulphur content is low, rarely above 1 per cent; ash content is very small. The principles governing the combustion of coal tar fuels are similar to those for fuel oils of comparable viscosity; oil-firing equipment for petroleum oils is generally suitable for coal tar fuels.

"Coalite". A reactive coke of the Coalite and Chemical Co. Ltd., produced under "low temperature" carbonization conditions when coal is heated in retorts to a temperature of the order of 1120° F (600° C). It ignites readily, burns with very little smoke emission, and is well-suited for all domestic appliances including old-fashioned stool-bottom grates. The volatile content is about 7 per cent with a bulk density of about 26 lb/ft³. It is an authorized fuel in a *smoke control area*, q.v. See **Authorized Fuels; Coke; "Rexco"**.

Cobalt 60. A radioactive isotope of the metallic *element*, q.v., cobalt, at. no. 27, at. wt. 58·933. It is usually produced by *neutron* irradiation in a *nuclear reactor*; it has a *half-life*, qq.v., of 5 years. Cobalt 60 is used as a gamma-ray source in the treatment of cancer and in industrial radiography.

Cochrane Abrasion Test. A test for the abradability or surface hardness of coke; the test consists in rotating 28 lb of coke, sized 2 to 3 in., in a drum for 55½ min at 18 rev/min. The product is then screened on a ⅛ in. square mesh screen; the percentage remaining on the screen being an index of surface hardness. An index is considered satisfactory if above 74.

Coefficient of Expansion. The increase in dimension of a material for each degree rise in temperature, expressed as a fraction of the original dimension; coefficients may relate to linear or cubic expansion. If a metal has a coefficient of linear expansion of 0·00002 per degC, this means that it expands by 0·00002 in. for every inch of its original length.

Coke. The solid residue remaining after the *destructive distillation*, q.v., of coal in ovens or retorts from which air has been excluded. The *volatile matter*, q.v., given off during the heating process is purified if for use as *town gas*, q.v. Coke has a high *fixed carbon*, q.v., content and a low volatile content (about 2 to 7 per cent). High temperature coke is difficult to ignite, whereas low temperature coke is relatively easy. The calorific value of coke is of the order of 12,500 Btu/lb (6,940 Chu/lb). See **Carbonization; Shatter Test.**

Coke Breeze. The residue from screening after all graded coke has been removed; it consists of pieces and particles below about a half-inch in size. A number of older power stations use coke breeze with coal on *chain grate stokers*, q.v.; the mixture may contain up to 15 per cent of coke.

Coke By-products. Important raw materials for the chemical industry obtained during the manufacture of coke; they include crude tar, crude benzole, ammonia liquor, toluene, xylene, naphthas and crude tar distillates. *Benzene*, derived from *benzole*, qq.v., is used to make nylon, dyestuffs, drugs and photographic chemicals. Toluene is used to make explosives, saccharine, dyestuffs, food preservatives, perfumes and drugs. Xylenes are used to make paint, varnish, rubber and printing ink. Naphthas are used to make waterproof material, metal and floor polishes, varnishes and black protective coatings. Tar distillates include road tar, tar fuel oil, pitch, creosote, tar acids, anthracene, naphthalene and pyridine.

Coke Oven. A retort used for the manufacture of metallurgical or hard coke, mainly for the iron and steel industry. A battery consists of 10 to 60 or more ovens; each oven is a slot-like chamber constructed of refractory brickwork, about 42 ft in length, 13 ft high and 16 in. in width. An oven of this size will carbonize 16 to 18 tons of coal in about as many hours. The oven ends are closed by self-sealing refractory-lined doors. Three or four charging holes, normally closed by heavy case iron covers, are situated in the oven roof. The refractory brickwork walls contain flues in which coke oven, blast furnace, or other gas is burned to heat the oven. The temperature of the oven reaches about 2370° F (1300° C). Each battery has one or two gas mains to remove the gas evolved in the ovens during carbonization; the connection between the oven and the main is known as an ascension pipe. A charging car runs along the top of the battery discharging coal into the ovens as required; another machine incorporates a ram for pushing the coke out of the oven.

Coke Residue. Synonymous with *fixed carbon*, q.v.

Coke Tree. A solid mass of coke formed by strongly caking coals in under-feed stokers; it grows as it is pushed upwards by fresh fuel fed from below. "Off" periods tend to aggravate the formation of coke trees and if manual breaking is resorted to, smoke is created.

Coker. An oil refining unit used to convert vacuum residuum to coke and distillates.

Coking. (a) An alternative name for "caking". See **Caking Coal.**
(b) The undesirable building-up of coke or carbon deposits in oil refinery equipment.

Coking Coal. *Bituminous coal*, q.v., suitable for the production of *coke*, q.v. See **Caking Coal.**

Coking or Deadplate Firing. A method, originally devised by James Watt, of firing a boiler or furnace with solid fuel. The coal is piled on the deadplate or on the front end of the bars to a depth of about 10 in. The heat of the furnace drives off the volatiles from the fresh coal. In passing over the incandescent firebed they ignite and burn. When the heap of fresh coal has been partially coked it is pushed forward over the grate. The method provides a most effective way of preventing smoke; it is emulated in the automatic coking stoker.

It does not produce as much steam per hour as the sprinkling and side-firing methods but, as refuelling is done at less frequent intervals, it is useful where the fireman has other duties to perform. More skill is required than with other methods to keep the whole grate covered and prevent *blow-holes*, q.v. The main problem is at the back of the grate which cannot be seen over the heap of fresh coal at the front. The rake must be frequently used. See **Firing by Hand.**

Coking Stoker. A *mechanical stoker*, q.v., for solid fuel in which the fuel is fed from a hopper by a ram on to a top coking or distribution plate at the front of and above the grate; the coal is thus partially carbonized by the heat of the furnace before spilling over on to the bottom coking or deadplate. The latter holds sufficient fuel to ensure the rapid ignition of the coal as it spills over from the top plate, thus creating a condition of both overfeed and underfeed ignition. The combination of the movement of the fuel bed caused by the reciprocating fire bars and the new coal pushed in over the top plate by the ram causes the ignited fuel to fall off the bottom coking plate and travel along the grate. There are two types of coking stoker, one the high ram, which is now considered obsolete, and the modern *low ram coking stoker*, q.v.

Cold End. That section of an *air preheater*, q.v., where the cooled flue gas leaves and cold air enters the heater.

Cold Gas Efficiency. In respect of a *gas producer*, q.v., thermal efficiency calculated as follows:

$$\text{Cold gas efficiency, } \% = \frac{\text{Potential heat of gas}}{\text{Total heat of fuel}} \times 100$$

where total heat means the sum of potential heat (calorific value) plus sensible heat due to preheating. Cold gas efficiencies range from 63 to 80 per cent. See **Hot Gas Efficiency.**

Cold Rolled Steel. Steel that is passed cold through a rolling mill.

Colloidal Fuel. Fuel oil containing 30 to 35 per cent of *pulverized fuel*, q.v; it is fired in the same manner as heavy fuel oil.

Colour/Temperature Scale. A scale applicable to hot metals at and above 1000° F (538° C) the temperature of which can be estimated by their colour. For iron and steel, the colour scale is broadly: dark red 1000° F (538° C); medium cherry red 1250° F (677° C); orange 1700° F (927° C); yellow 1850° F (1010° C); white 2400° F (1316° C).

Combustion. A state of chemical activity in which the reactive elements of a fuel burn or unite with oxygen, accompanied by the evolution of heat and often light. Complete combustion means, in effect, oxidization to the highest possible degree. One pound of carbon, during complete combustion to carbon dioxide, produces about 14,500 Btu. If hydrogen burns to water vapour about 62,000 Btu are produced per pound of hydrogen burned. The heat liberated in the complete combustion of sulphur to sulphur dioxide is about 4000 Btu for every pound of sulphur burned. However, before most

TABLE 2—STAGES OF COMBUSTION OF COAL

Stage 1	Carbonization (release of volatiles; production of H_2 and CH_4; formation of coke). Rate dependent upon: (a) heat input; (b) heat capacity; (c) thermal diffusivity; (d) particle size
Stage 2	Partial oxidation of coke and volatiles, together with pyrolysis (polymerization) of volatiles, leading to formation of: (a) CO; (b) H_2; (c) hydrocarbons; (d) carbon
Stage 3	Combustion of: (a) H_2 and CO; (b) hydrocarbons (via CO and H_2); (c) carbon and coke (via CO) $\rightarrow H_2O + CO_2$.
Stage 4	Gasification reactions: (a) $C + H_2O + CO_2$ (b) Hydrocarbons $+ H_2O + CO_2 \rightarrow H_2 + CO \rightarrow H_2O + CO_2$

(Adapted from "Twelfth Coal Science Lecture", D. T. A. Townend, *BCURA Gazette*, No. 48, 1963.)

of the heat available from coal can be released with the formation of the ultimate oxidation products CO_2 and H_2O, the fuel has to be converted first into CO and H_2; this is true for all hydrocarbons. Before combustion can take place an initiating "ignition temperature" must be attained. The four stages of combustion of coal are illustrated in Table 2.

Combustion Chamber. Any enclosed, or partially enclosed, space in which the processes of *combustion*, q.v., take place. The process of combustion may not be completed, however, within the confines of a single chamber and two or more chambers may be employed. In a two-chamber arrangement, the primary chamber is where ignition and burning of the fuel or waste takes place; in the secondary chamber the burning of vapours, gases and particulate matter from the primary chamber is completed.

Combustion, Chemical Reactions of. The chemical combination of the fuel elements or constituents with oxygen; the most common basic equations for combustion are as shown in Table 3.

TABLE 3—CHEMICAL REACTIONS OF COMBUSTION

Fuel	Reaction (Heat of reaction is ignored)
Carbon (to CO)	$2C + O_2 \rightarrow 2CO$
Carbon (to CO_2)	$C + O_2 \rightarrow CO_2$
Carbon monoxide	$2CO + O_2 \rightarrow 2CO_2$
Hydrogen	$2H_2 + O_2 \rightarrow 2H_2O$
Sulphur (to SO_2)	$S + O_2 \rightarrow SO_2$
Sulphur (to SO_3)	$2S + 3O_2 \rightarrow 2SO_3$
Hydrogen sulphide	$2H_2S + 3O_2 \rightarrow 2SO_2 + 2H_2O$
Acetylene	$2C_2H_2 + 5O_2 \rightarrow 4CO_2 + 2H_2O$
Ethane	$2C_2H_6 + 7O_2 \rightarrow 4CO_2 + 6H_2O$
Ethylene	$C_2H_4 + 3O_2 \rightarrow 2CO_2 + 2H_2O$
Methane	$CH_4 + 2O_2 \rightarrow CO_2 + 2H_2O$

Combustion Meter. A combined instrument for the measurement of steam flow and of air flow. A recorder enables the *steam flow/air flow ratio*, q.v., to be clearly seen; adjustments are usually made so that the recording pens coincide when the desired combustion conditions are obtained.

Comminution. Pulverization; the reduction of material to a powder by attrition, impact, crushing, grinding or by a chemical method.

Common Banded Coal. Term used in the United States of America for common varieties of bituminous and sub-bituminous coals.

Common Rail System. A category of *fuel injection equipment*, q.v., for diesel engines; the injection nozzles to each cylinder are supplied by a common fuel line, a fuel pump maintaining a pressure of from 4000 to 10,000 lb/in² in the line. The fuel injection valves open and close mechanically. The *Jerk System*, q.v. is now more common. See **Diesel Engine.**

Commutator. A device which reverses the connections from the load-circuit to the rotating coils at the instant when the electromotive force (e.m.f.) in the coils is changing its polarity; it is added to an alternator to obtain direct current (d.c.).

Compound Turbine. A *steam turbine*, q.v., in which steam is expanded in a number of separate cylinders. In a "tandem compound turbine" the cylinders are in line and the rotors coupled; in a "cross-compound turbine" the cylinders are in two lines driving two generators.

Compression Ratio. The ratio of the volume of the fuel-air mixture at the start of compression to the volume of the fuel-air mixture at the end of compression, in the cylinders of an internal-combustion engine.

Computer. A device capable of carrying out the following five main functions: (a) acceptance of information; (b) storage of information; (c) simple mathematical operations such as add, compare, etc.; (d) output of information in the form required; and (e) control of its own operation. Computers are grouped into two main types: (a) those using analogue techniques; (b) those using digital techniques. See **Analogue Computer; Digital Computer.**

Condensate. Steam which has given up its latent heat and changed back again into water. Condensate forms a film on internal metal surfaces, running down such surfaces to a draw-off point at a lower level.

Condenser. A chamber in which exhaust steam from, say, a *steam engine* or *steam turbine*, qq.v., is condensed or converted to water by the circulation or introduction of cooling water. See **Jet Condenser; Regenerative Condenser; Surface Condenser.**

Condensing Turbine. A *steam turbine*, qq.v., in which all the steam used to drive it is finally expanded down to the condenser pressure which may be a vacuum of between 28·5 and 29·0 in. of mercury.

Conditioning. The addition of water or steam to dusty coals or fines to improve the distribution of air through the fuel bed and reduce the quantity of dust carried forward towards the chimney.

Conductance. The reciprocal of resistance, being measured in mhos (or reciprocal ohms); a 5 ohm resistor has a conductance of 0·2 mhos.

Conduction. The transmission of heat from one part of a substance to a colder part, or to another substance or body in contact with it. The heat transfer rate may be determined by the following formula:

$$H = \frac{K(t_1 - t_2)}{L}$$

where, H = heat transmission in Btu/ft² h;
   L = thickness of the substance through which the heat passes, in inches;
   t_1 = temperature of the hotter face in ° F;
   t_2 = temperature of the cooler face in ° F;
   K = thermal conductivity of the substance expressed as Btu/ft² h ° F 1 in. thickness.

Examples of the constant K are: copper, 2500; wrought iron, 416; silica brick, 8 to 10; insulating brick, 0·6 to 2·0; diatomaceous brick 0·6 to 0·8. See **Heat Transfer.**

Coning. The behaviour of a chimney plume when it expands or diffuses roughly along a cone; during such periods *efflux velocity*, q.v., or momentum and thermal buoyancy tend to be dominant. See **Looping.**

Conradson Test. A test to determine the *carbon residue*, q.v., of a liquid fuel. In this test 10 g oil are heated in a crucible by a gas burner at such a rate that oil vapours cease to be generated after 13 min.; the temperature is then raised to full redness for a further 7 min., and then the crucible is allowed to cool. The residue is weighed and expressed as a percentage of the oil used. See **Ramsbottom Test.**

Constant Volume Gas Thermometer. A device for measuring temperature. It comprises a bulb containing an inert gas and a measuring element connected by capillary tubing. The measuring device registers the change in pressure of a constant volume gas, the pressure varying with changes in temperature. It is a useful instrument for measuring low temperatures and may be used up to 1022° F (550° C). See **Pyrometer; Thermometer.**

Constitutional Formula. A chemical formula which indicates the actual grouping of the atoms in the molecule. For example, normal butane with an *empirical formula* of C_2H_5 and a *molecular formula*, qq.v., of C_4H_{10} has a constitutional formula of $CH_3(CH_2)_2CH_3$. See **Graphical Formula.**

Continental Shelf Act, 1964. An Act which allowed the Ministry of Power to issue exploration and production licences for the British Sector of the North Sea Continental Shelf.

Continuous Blow-down. The draining of a constant stream of water from a boiler, often through a feed water heat exchanger. See **Blow-down.**

Continuous Vertical Retort. A closed chamber used for the manufacture of *town gas*, by the *carbonization* qq.v., of coal. A retort of this type is usually about 22 ft high and 10 ft wide; the depth varies from 8 in. at the top to 12 in. at the bottom. It holds about $3\frac{1}{2}$ tons of coal. The retort is heated by producer gas burnt at the sides. Coal is fed continuously into the retorts through gas-tight hoppers, coke being extracted continuously, followed by "steaming" in an air-tight coke box. This steam quenches the coke and produces *water gas*, q.v., which increases the total thermal yield of gas per ton of coal carbonized. While the coke is ejected continuously into the coke box, the coke box itself is emptied periodically. The steamed vertical retort produces a gas with the following typical composition, the constituents being expressed in percentages: Hydrogen, 50; methane, 20; carbon monoxide, 15; nitrogen, 7; carbon dioxide, 5; hydrocarbons, 2; oxygen, 1.

Control Rod. A rod, consisting usually of steel containing a good neutron absorber such as *boron* or *cadmium*, qq.v., which is used to control the reactivity of a *nuclear reactor*, q.v. Movement of the control rods enables the power level to be held constant or varied.

Convection. The transmission of heat from one place to another through the medium of a heat-transporting fluid. Convection in still air may be calculated from the formula:

$$Hc = C(t_1 - t_2)^{1.25}$$

where, $Hc = \text{Btu/ft}^2 \text{ h}$

$t_1 = $ temperature of surface of the heat source in ° F;

$t_2 = $ temperature of heat-transporting fluid in ° F;

$C = $ a constant, being 0·3 for vertical upright surfaces, 0·2 for horizontal surfaces facing upwards, and 0·35 for pipes and large horizontal cylinders.

See **Heat Transfer.**

Convector Superheater. See **Superheater.**

Converter Reactor. A *nuclear reactor*, q.v., using one kind of fissile fuel and producing a different kind of fissile material; for example the production of Pu^{239} in a reactor using U^{235}.

Converting. The process whereby neutrons are used to transmute thorium 232 into uranium 233, or uranium 238 into plutonium 239.

Coolant. A heat-transport medium for removing heat from a *nuclear reactor*, q.v., so that it may be used for steam raising and power production. Among the more frequently used reactor coolants are ordinary (light) water, heavy water, carbon dioxide and liquid sodium. British commercial nuclear reactors have employed carbon dioxide as the coolant; this gas is used under pressure. Pressures used at various U.K. nuclear power stations in lb/in^2 are, Berkeley, 125, Bradwell, 132, Hinkley Point, 185, Trawsfynydd, 239; Dungeness 'A', 286; Dungeness 'B', 450.

Cool Flame. Flame produced by *hydrocarbon*, q.v., mixtures at 390 to 750° F (200 to 400° C) at pressures slightly above atmospheric pressure; combustion is incomplete and the products of combustion include aldehydes, ethers and peroxides. Cool flames are blue in colour and slow moving. See **Flame**.

Cooling Tower Precipitation. The drizzle formerly experienced around natural draught cooling towers; this nuisance has been overcome by the general use of spray eliminators fitted to the inside of towers.

Copper Sweetening Unit. An oil refinery unit in which corrosive and unpleasant compounds in motor spirit are converted in order to give a non-corrosive product with a pleasant odour.

Core. The part of a *nuclear reactor*, q.v., containing the fissile material.

Corner-fired Furnace. A *water-tube boiler*, q.v., roughly square in plan in which the pulverized fuel burners are arranged in vertical banks at each corner, and directed at points to one side of the centre line of the furnace. This results in the formation of a large vortex with its axis on the vertical centre line; the effect is to promote a high degree of turbulence. The burners consist of an arrangement of slots one above the other, admitting through alternate slots primary air/fuel mixture and secondary air. It is usually possible to tilt the burners upwards or downwards with a maximum inclination to the horizontal of 30°. Thus the position of the main flame region in the furnace may be varied, affecting furnace outlet temperatures and permitting superheat control. Corner-fired furnaces are used with medium and high volatile bituminous coals. See Fig. 4.

Cornish Boiler. A horizontal shell boiler of simple design contain-

59

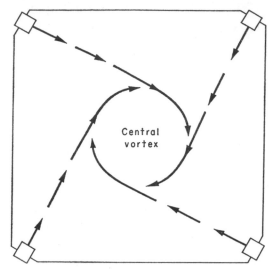

Central
vortex

FIG. 4. *Horizontal section of corner-fired water-tube boiler furnace,*
showing general flow pattern.

ing a single furnace tube. It has a brick flue setting similar to that of
the Lancashire boiler. The hot combustion gases leaving the grate
pass to the back end of the single furnace tube and down into a
bottom flue immediately below the boiler; the gases return along
this flue to the boiler front and then divide into two streams passing
down side flues, one being situated on each side of the boiler, to the
rear. Here the gases re-unite in a single main flue and pass to the
chimney. Evaporative capacities range from about 2000 to 4500
lb/h "from and at 212° F", q.v. See **Boiler; Lancashire Boiler.**

Corona. A luminous discharge, present in the *electrostatic precipitator*, q.v., due to air break-down when the electric stress at the
surface of an electrode exceeds a certain value.

Coulomb. A unit of quantity of electricity; it is the quantity of
electricity transported in one second by a current of one *ampere*,
q.v.

Creep. The continuous deformation of metals under steady load.
At elevated temperatures, creep is exhibited by iron, copper, nickel
and alloys; at room temperatures, creep is exhibited by such metals
as lead, tin and zinc. See **Creep Test.**

Creep Test. A method of measuring the resistance of metals to *creep*, q.v.; a test piece of metal is subjected to a fixed static tensile load, the temperature being kept constant throughout the test. The elongation of the test specimen is measured accurately at regular intervals, the results being plotted on a length/time graph. Creep is divided into primary, secondary and tertiary periods, each period having its own characteristic. There are many variations of this test in terms of duration and procedure adopted.

Critical. The condition of a *nuclear reactor*, q.v., which is just capable of sustaining a chain reaction. See **Critical Mass.**

Critical Air Blast (C.A.B.) Test. A laboratory test to determine the *reactivity*, q.v., of solid fuels, particularly cokes; it determines the minimum rate of air supply, in ft^3/min, sufficient to keep a standard bed of 14 to 35 B.S. mesh fuel alight. The lower the C.A.B. value, the more reactive the fuel. Cokes having C.A.B. values below 0·050 are readily lighted and maintained in approved open-fire grates.

Critical Mass. The minimum mass of fissile material needed for a chain reaction of neutrons to be self-sustaining. See **Critical.**

Critical Pressure. Steam pressure below which latent heat is required to convert a liquid into vapour.

Critical Pressure Gauge. A steam pressure gauge which gives a high reading accuracy; full scale deflection of the pointer follows a change in pressure of only a few lb/in^2 variation from the selected pressure.

Critical Temperature. Temperature above which it is impossible to liquify a gas by pressure alone.

Cross-Channel Cable. A cable connecting the electricity systems of Britain and France. As the times of maximum load in each country do not coincide electricity can be sent from one country to the other as may be economically desirable. For this purpose, *alternating current* is converted to *direct current*, qq.v., and re-converted to alternating current at the receiving end.

Cross-section, Total. The sum of the cross-sections of a nucleus which comprise: (a) capture cross-section—the apparent area of a nucleus for capture as "seen" by the neutron; (b) scatter cross-section—indicating the probability of a neutron colliding with a nucleus and rebounding without altering the structure of the nucleus; (c) fission cross-section—which indicates the probability of a neutron causing fission. See **Barn; Neutron; Nucleus.**

Crude Oil. Petroleum as it occurs naturally in the earth consisting essentially of a mixture of gaseous, liquid and solid hydrocarbons; sulphur, oxygen, nitrogen, derivatives of hydrocarbons and traces of inorganic elements are also present. *Natural gas*, q.v., a mixture of gaseous hydrocarbons, is normally present with the crude oil, accumulating in the upper parts of the oil bearing strata. While crude oil occurs throughout the world, 90 per cent of production is mainly in North, Central and South America, the Middle East and the USSR. The composition of crude oil from all sources falls generally within the following ranges, the constituents being expressed as percentages: carbon, 83 to 87; hydrogen, 11 to 15; sulphur, 0·1 to 6·0; nitrogen, 0·1 to 1·5; oxygen, 0·3 to 1·2. See **Petroleum.**

Crude Oil, U.S. Bureau of Mines Classification of. A classification of petroleum crudes into seven categories, according to their "base", from the distillation analysis of 800 samples of *crude oil*, q.v., from all over the world. The categories are:

Class A; Paraffin base oil (wax-bearing)
Class B; Paraffin intermediate base oil (wax-bearing)
Class C; Intermediate paraffin base oil (wax-bearing)
Class D; Intermediate base oil (wax-bearing)
Class E; Intermediate naphthene base oil (wax-bearing)
Class F; Naphthene intermediate base oil (wax-bearing)
Class G; Naphthene base oil (wax-free).

Crude Unit. An oil refining unit comprizing an atmospheric or low pressure first stage distillation unit, usually followed by a second stage distillation unit operating under vacuum. The complete unit comprises *distillation towers, heat exchangers* qq.v. and heaters.

Cupola. A vertical shaft metal-melting furnace comprizing a cylindrical mild steel shell lined with refractories; "drop bottom" doors are fitted to the base. A tap-hole for the draining of molten metal is situated at the lowest point; above this on the opposite side is another tap-hole for the removal of slag formed during melting. Tuyeres for the admission of an air blast are situated between 2 to 4 ft from the working bottom. Cupolas vary in height from about 10 ft up to about 70 ft. The cupola is charged with alternate lays of metal, coke and flux introduced through a charging hole high up in the shell. The output of a cupola is roughly proportional to the cross-sectional area of the melting zone—about 0·4 to 0·8 tons/h ft^2; total outputs range up to 30 tons/h. Metal to coke ratios vary from 5 : 1 to 12 : 1. The air blast is normally cold, but an increasing number of hot-blast cupolas are coming into use.

Curie. (c). The unit of radioactivity. It is the quantity of any radioactive nuclide in which the number of disintegrations per second is 3.7×10^{10}. Sub-units are the millicurie (mc) $= 10^{-3}$c and the microcurie (μc) $= 10^{-6}$c.

Current. The movement or "flow" of electricity: usually measured in amperes.

Curtain Wall. A refractory construction or baffle which serves to deflect combustion gases in a downward direction. Also known as a "drop arch".

Cut out. A fuse such as the main fuse provided by the Electricity Board on a consumer's premises.

Cut Point. A specified value in respect of, say, density or size, at which a separation into two fractions takes place.

Cuts. Fractions obtained in the distillation of oil.

Cycle Efficiency. The *adiabatic heat drop*, q.v., expressed as a percentage of the heat put into the steam. See *Rankine Cycle*.

Cyclone. (a) A device for removing particulate matter from the waste gases of industrial processes. A simple cyclone consists of a cylindrical upper section and a conical bottom section. The dust laden gases enter the cylindrical section tangentially. Centrifugal action throws the grit and dust particles to the outer walls; these particles fall by gravity to the dust outlet at the bottom. The relatively clean gases leave through a centrally situated tube within the upper section. With a pressure drop of 3 in. w.g. separation of particles down to to about 40μ can be achieved. Cyclones may be used singly, or in groups or nests. (b) In meteorology, a low pressure area with winds rotating counter-clockwise around the centre in the northern hemisphere, and clockwise in the southern hemisphere. Winds bring in moisture and rainy, windy weather prevails as rising air cools and vapour condenses. See **Anticyclone.**

Cyclone Furnace. A furnace in which crushed coal is burnt in a water-cooled cyclone, the hot gases passing into a secondary furnace in which the grits and semi-molten ash are trapped before the gases continue into the boiler. See Figure 5.

Cyclone Gasifier. A gasification chamber in which the blast is admitted tangentially to produce a swirl.

Cyclone Washer. A coal-washing device which enables very small coal to be washed with a precision virtually as high as that obtained using the *dense medium process*, q.v. The cyclone washer does not

Cyclone Washer

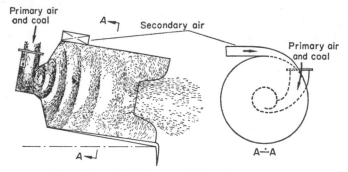

FIG. 5. *Simplified diagram of the cyclone furnace.*

depend on gravity alone to effect the separation of coal and dirt, but utilizes centrifugal force; a cyclone of 20 in. diameter is capable of washing up to 50 tons of small coal per hour.

Cylinder Displacement. The volume of the compression space in the cylinder of an internal-combustion engine at the beginning of the compression stroke of the piston, less the compression space at the end of compression; or the cross-sectional area of the cylinder multiplied by the length of the stroke.

Cylindrical Screen. A screen revolving at 3–4 rev/min on rollers; as coal passes through it, the smalls pass through the holes.

Cylinder Stock. The *residuum* remaining in a *still*, qq.v., after the lighter parts of the crude have been vaporized.

D

Damper. A manual or automatic device to regulate the rate of flow of combustion gases through the flues of a fuel-burning appliance. See **Barometric Damper; Butterfly Damper; Guillotine Damper; Sliding Damper.**

Dark Smoke. Commonly defined as smoke appearing to be as dark as, or darker than, shade 2 on the *Ringelmann Chart*, q.v.

Data Logger. An automatic device which records events and physical conditions, usually with respect to time; it is employed where a large number of plant parameters such as pressure, temperature or flow, have to be measured, displayed and analyzed. The results of the analysis are used to produce a more efficient operation of the plant or to detect abnormal conditions that may arise during the normal day-to-day running of the plant. The system usually employs a single measuring instrument which is shared between the many inputs; it may be of analogue or digital type. A data logger does not normally have storage facilities for processed information See **Computer.**

Daughter. A *nuclide*, q.v., formed by the disintegration of another nuclide called the *parent*, q.v.

Dayton Process. A process for the manufacture of gas from oil; it is a continuous and non-catalytic process. Vaporized distillate is partially combusted with air at 1300 to 1650 °F (704 to 899° C); the air used is about 10 per cent of that required for stoichiometric combustion. The constituents of a typical gas produced, expressed in percentages, are: hydrocarbons, 25; carbon monoxide, 6; carbon dioxide, 6; hydrogen, 2; nitrogen, 61. The calorific value is in the range 400 to 500 Btu/ft^3 (222 to 278 Chu/ft^3). See **Dayton-Faber Process.**

Dayton-Faber Process. A process for the manufacture of gas from oil; it is a development of the *Dayton Process*, q.v., using oxygen-enriched air in a modified form of generator. The calorific value is in the range 500 to 900 Btu/ft^3 (278 to 500 Chu/ft^3)

Dead Hearth. A solid hearth; one through which no combustion air passes.

Dead Plate. A cast-iron plate bolted to the front of a boiler furnace tube and bevelled to carry fixed firebars; no combustion air passes through it.

"Dead Plate Firing". See **Coking or Deadplate Firing.**

Dead Storage. Storage arrangements in which some fuel tends to be never recovered; this can occur with badly designed bunkers. Bunkers constructed with angled sides will prevent dead storage, and ensure continuous movement and passage of solid fuel. The minimum angles of hopper sides, measured from the horizontal, should be for steel and concrete surfaces respectively (a) coke, 45° and 50°; (b) graded coal, 40° and 47½°; (c) washed small coal, 55° and 60°.

Dead-weight Safety Valve. A *safety valve*, q.v., the operation of which depends upon a series of weights superimposed upon it; the blow pressure can be varied by adding or removing individual weights.

Dead-weight Tons. The total carrying capacity of a ship when loaded to the appropriate freeboard. Dead-weight tonnage is very roughly about one and a half times the gross registered tonnage (which is in "tons" of 100 cubic feet).

De-aerator. Plant for removing the *dissolved gases*, q.v., particularly oxygen, from feed water.

Dean and Stark Test. A standard test for determining the water content of a fuel oil.

Decarburizing. The removal of carbon from iron carbide (Fe_3C), thus affecting the crystalline structure of the steel. The presence of carbon dioxide can decarburize steel. See **Carburizing.**

Degrees A.P.I. A scale indicating the gravity of oil, adopted by the American Petroleum Institute. The formula is:

$$\text{Degrees A.P.I.} = \frac{141 \cdot 5}{\text{Sp. gr. } 60° \text{ F}/60° \text{ F}} - 131 \cdot 5$$

Delayed-coking Process. An oil refinery process in which residual oil is heated and pumped to a reactor to coke; the coke is deposited as a solid mass of granular material and is subsequently broken up into lumps. See **Fluid-coke Process; Petroleum Coke.**

Deliquescence. The absorption by a substance of moisture upon exposure to the atmosphere. Substances that are ordinarily deliquescent are concentrated sulphuric acid, glycerol, calcium chloride crystals, sodium hydroxide solid, and ethyl alcohol 100 per cent. In an enclosed space these substances deplete the water vapour present to a definite degree.

Dense Medium Process. A wet cleaning process for separating "dirt" in the form of sandstone, shale, clay, pyrites, calcite and so on from *run-of-mine coal*, q.v. In this process an aqueous suspension of finely ground heavy solids, e.g. magnetite or barytes, is used to effect separation; *coal, middlings*, qq.v., and shale can be separated with a high degree of efficiency. Separation may be achieved in two stages; in one bath with a specific gravity of 1·4 the clean coal is floated and removed, while in the second bath with a specific gravity of 1·8 the middlings float while the shale sinks.

Density. The weight of a substance or gas in relation to its volume. The weight of 1 cubic foot of water at 39·2° F (4° C) and normal atmospheric pressure is 62·5 lb, i.e., the density of water is 62·5 lb/ft^3. The temperature and pressure at which measurements are made must always be stated. Since equal volumes of gas contain the same number of molecules, the molecular weights of all gases, expressed in the same units of weight, occupy the same volume. It

TABLE 4—DENSITIES OF VARIOUS GASES AT N.T.P.

Gas	$d,g/l$	$d,lb/ft^3$
Air	1·2929	0·08071
H_2	0·08988	0·005611
O_2	1·42904	0·08921
N_2	1·25055	0·07807
CO_2	1·9769	0·12341
SO_2	2·9269	0·1827
H_2O	0·5980	0·0373
(Steam 100°)		
CO	1·2504	0·07806
CH_4	0·7168	0·04475
H_2S	1·539	0·09608
NH_3	0·7710	0·04813
CL_2	3·214	0·2006
NO	1·3402	0·08367
N_2O	1·9778	0·1235

Adapted from, *American Institute of Physics Handbook: Spiers' Technical Data on Fuel*

has been determined experimentally that within certain limits the molecular weight of any gas expressed:

(a) in grammes, occupies 22·4 litres at 32° F (0° C) and 760 mmHg pressure

(b) in pounds, occupies 359 ft³ at 32° F (0° C) and 30 inHg or 379 ft³ at 60° F (15·6° C),

the gas being dry in every case. These amounts are known as the gramme-molecular weight and the pound-molecular weight, respectively. If the gas is saturated with water vapour at 60° F (15·6° C) the volume occupied at a total pressure of 30 inHg is 385 ft³. The density of a number of gases is given in Table 4. The density at any temperature and pressure can be obtained from the expression:

$$d = d\left(\frac{P}{760}\right)\left(\frac{273}{T}\right)$$

where P is the barometric pressure in millimetres of mercury, and T is the absolute temperature in degrees Kelvin. See **Specific Gravity.**

Derv. Abbreviation of "diesel engined road vehicle", a term still widely used in Britain for distillate fuel suitable for high-speed diesel engines.

Destructive Distillation. A distillation process in which an organic compound or mixture is heated to a temperature high enough to cause decomposition. See **Charcoal.**

Desuperheater. See **Attemperator.**

Detergent. A lubricating oil additive intended to prevent the formation of deposits in internal combustion engines; typical compounds are organo-metallic substances such as phenates and sulphonates of barium and calcium.

Detonation Meter. An instrument for measuring knock intensity, detecting the rate of change of pressure in an engine cylinder. See **Knocking.**

Detoxification. The reduction of the *carbon monoxide*, content of *town gas*, qq.v.

Deuterium. $_1H^2$. The hydrogen isotope, mass number 2; commonly known as "heavy hydrogen", it is obtained from heavy water. It occurs in natural hydrogen in the proportion 1 : 6500. See **Deuteron.**

Deuteron. The nucleus of the deuterium atom, consisting of a proton and a neutron. See **Deuterium.**

Deutsch Formula. A formula devised to calculate the performance efficiency of an *electrostatic precipitator*, q.v.:

$$\text{Efficiency} = 1 - \exp(-FC)$$

where, F = specific collecting surface (projected collecting electrode surface area per unit volume of gas treated per second);
C = effective migration velocity (e.m.v.) of the particles.

Any self-consistent system of units may be used in the above equation, although e.m.v. is usually expressed in centimetres per second. E.m.v. depends mainly on particle size and on the strength of the electric field between the electrodes.

Dew Point. The temperature at which moisture in air or flue gas condenses. If flue gases contain little or no sulphur trioxide then the dew point of the gases is known as water dew point; this is usually in the region of 120° F (49° C). If, however, significant amounts of sulphur trioxide are present the dew point may be raised as high as 300° F (149° C); this is known as the acid dew point. See **Acid Soot.**

Dewaxing. The removal of wax from lubricating oils; wax tends to crystallize out at low temperatures and impair operation, although a small amount is desirable for its own lubricating qualities. The amount permitted to remain in an oil depends on the operating conditions envisaged for that oil.

Diatomaceous-earth-base Blocks. An insulating material consisting of silica, uncalcined or calcined, with asbestos fibre and clay, moulded under heat and pressure. Used mainly for boiler walls behind firebrick in temperature zones, it is claimed, up to 1800° F (982° C).

Dielectric Heating. The production of heat in non-conducting materials by their losses when subject to a high frequency alternating electric field. Frequencies required range from 1 to 200 Mc/s. The non-conducting material to be heated is placed between two electrodes across which the high-frequency voltage is applied.

Diesel Engine. An internal combustion engine in which air is compressed, attaining a high temperature, followed by the injection of oil which ignites; there is no carburettor or spark ignition system. Each piston in a four-stroke engine moves up and down twice to produce a power impulse:

(a) Intake stroke—the piston descends drawing in air only;

(b) Compression stroke—the piston ascends compressing the air to about one-fifteenth of its original volume raising its temperature to about 850° F (455° C);

(c) Power stroke—near the end of the compression stroke a very accurately measured quantity of oil fuel is injected which ignites immediately, the expansion of the gases forcing the piston down;

(d) Exhaust stroke—the piston ascends and the burnt gases are discharged.

The air used is unthrottled when the intake valve opens and a high compression ratio is employed to heat the air sufficiently to ignite the fuel (a higher ratio than with a petrol engine). The rate of fuel injection is varied to change the power output. With the high air/fuel ratio diesel exhaust gas contains much less carbon monoxide and hydrocarbon than the gasoline engine, but somewhat more *nitrogen oxides*, q.v., and aldehydes. Engine overloading and poor maintenance can be the causes of smoke emission; basically there should be none. The thermal efficiency of a diesel engine varies between 25 to 33 per cent. See **Fuel Injection Equipment; Gasoline Engine; Motor Vehicle Exhaust Gases.**

Diesel Index. An estimation of ignition quality based on the aniline point and the specific gravity of a diesel fuel:

$$\text{Diesel index} = \frac{\text{Aniline point }°F \times \text{A.P.I. Gravity}}{100}$$

Diesel indices are generally about three numbers higher than the corresponding *cetane number*, q.v., for fuels with cetane numbers of about 45 to 50, but this can vary considerably. The ignition quality of diesel fuels can be improved by using various additives which promote the oxidation mechanism of the fuels. Some of the additives used are: (a) alkyl nitrates; (b) aldehydes, ketones, ethers, esters and alcohols; (c) peroxides; (d) aromatic nitro compounds. See **Degrees A.P.I.**

Diesel Locomotive. A form of railway traction utilizing the *diesel engine*, q.v.; it offers power, acceleration, efficiency and cleanliness. Diesel locomotives are of two types: (a) diesel-electric locomotives in which a diesel driven electric-generator supplies power to an electric motor drive; (b) diesel hydraulic locomotives in which the diesel engines supply power to the wheels through a torque convertor in the transmission line.

Diesel Oil. A distillate fuel with a typical specification as follows: (a) viscosity, Redwood No. 1 at 100° F (38° C), 34 seconds; (b) flash point, 170° F (77° C); (c) distillation range 356 to 680° F (180 to 360° C); (d) carbon residue, 0·02 per cent; (e) pour point, 10° F (−12·2° C); (f) sulphur, 0·5 to 1·0 per cent; (g) sp. gr. 0·84; (h) gross calorific value, 19,600 Btu (10,885 Chu).

Differential Pressure Meter. A meter equally suitable for measuring water from the delivery side of a feed pump and for measuring steam. It consists of a pressure generating element fitted in the main, a meter body in which the pressure difference is converted to a movement of an indicator by means of a float in a mercury-filled U-tube, together with an indicator, *recorder*, and *integrator*, qq.v. The differential pressure generating element consists of an orifice plate or venturi tube inserted in the steam or water main.

Diffusion. The spreading or scattering of a gas or liquid, or of heat and light. The spreading or intermixing movement of gaseous or liquid substances is due to: (a) molecular movement; (b) turbulence. In air pollution studies of the general atmosphere molecular diffusion is ignored, its effect being insignificant when compared with that of turbulence.

Diffusion Flame. A long luminous flame sustaining a virtually constant rate of radiation throughout its designed length of travel; diffusion occurs between adjacent strata of air and gas.

Diffusion Flame Burner. A *gas burner*, q.v., in which all the combustion air is provided by low-velocity diffusion in the combustion chamber.

Digital Computer. A *computer*, q.v., or fast automatic calculator in which data is represented by means of digits, i.e., the designation of one out of a finite number of alternatives by a digit or a group of digits. A digital computer consists of five basic units: (a) a central arithmetic unit where all calculations are carried out; (b) an input unit which enables all data to be read into the computer; (c) an output unit to read out the results of calculations from the computer; (d) a storage unit to store all data read into the computer, the results of calculations carried out within the computer and the instructions required by the computer to carry out the calculations; (e) a control unit which obeys the instructions written into the computer storage unit and directs the flow of information within the computer to perform the required computation. The instructions are called the program, the computer operating in a pre-determined manner as specified by this program. The problems best solved by digital computers are those in which: (a) large sets of algebraic equations have to be solved; (b) large quantities of numerical data are involved; (c) many logical decisions are demanded; (d) very high orders of solution accuracy are required. Digital computers are being used in a wide variety of industries for direct control of plant and machinery, e.g., in chemical and petroleum processes, in steel mills, and in power stations both conventional and nuclear. Off-line applications have included studies of coal allocation between power stations, short-term load predictions, calculation of ground level concentrations of sulphur dioxide, transport problems, regression analyses, thermal stresses in piping systems, vibration of turbine blades, temperature distribution in a cylindrical body, stress analyses and solutions of equations, and many others.

Diphenyl Air Heater. An air heater in which the heat-conducting medium is diphenyl oxide or other suitable liquid. The liquid is heated as it passes through a *heat exchanger*, q.v., situated in the flue gas stream, and cooled as it passes through another heat exchanger in the air stream.

Direct Burner. A *burner*, q.v., in which the fuel and oxidizer are mixed at the point of ignition. See **Premix Burner**.

71

Direct Current

Direct Current (d.c.). Electrical current flowing in one direction only. See **Alternating Current.**

Direct-fired Combustion Equipment. Plant in which the flame and/or the products of combustion come into direct contact with the material being processed, as in *brick kilns* and *open-hearth furnaces*, qq.v. See **Indirect-fired Combustion Equipment.**

Discharge Lamp. A form of electric lamp used extensively for street lighting, and in industrial premises. In the discharge lamp the current passes through a special mixture of gas and metallic vapour, forming a luminous electric discharge. The light may be greenish (mercury) or yellow (sodium) but in amount there is from two and a half to five times as much available as from filament lamps of equivalent consumption.

Distintegration. A nuclear transformation characterized by the emission of energy in the form of particles or photons. See **Curie.**

Distintegration Energy. (Q). The energy evolved, or absorbed, in a nuclear disintegration. It is equal to the energy equivalent of the change of mass which occurs in the reaction. If Q is positive, the disintegration is *exothermic*, q.v.; if Q is negative, it is *endothermic*, q.v. Radioactive disintegrations all have positive Q values.

Dispersoid. A colloidal or finely-divided substance.

Displacement Pump. A piston type pump used for boiler feed purposes. The commonest type of displacement pump is the single barrel simplex pump; the duplex pump has two water barrels which ensure a more uniform flow of water. The steam consumption of these direct acting pumps varies between 2 and 5 per cent of the boiler output. They are manufactured in a range of sizes discharging up to 20,000 lb water/h. They are reliable pumps, being mainly used where boiler pressures do not exceed about 200 lb/in^2.

Dissociation. A phenomenon which occurs at high temperatures when carbon and hydrogen, after combining with their full complement of oxygen, are broken up into molecules of fuel and oxygen, and molecules of oxygen. For example, an atom of carbon combines with two atoms of oxygen to form CO_2; when the furnace temperature rises to over about 2790° F (1530° C) the CO_2 may break up into carbon monoxide and oxygen, and if the temperature further increases to about 3000° F (1650° C) the CO molecule may break up into carbon and oxygen. The process is endothermic. See **Endothermic Reaction.**

Dissolved Gases. Undesirable gases in feed water such as oxygen,

72

carbon dioxide and ammonia; they may cause corrosion to feed pipes, feed-heater tubes and pumps. See **De-aerator.**

Dissolved Solids. Dissolved mineral salts in water supplies; in solution these salts divide almost entirely into their component parts, known as ions, which carry an electrostatic charge having one or more electrons too many or too few. The positively charged ions, called cations, include calcium, magnesium, sodium, and potassium. Negatively charged ions, called anions, include bicarbonate, carbonate, chloride, hydroxide, nitrate and sulphate. Dissolved solids increase in concentration as water is evaporated into steam and tend to promote foaming in the boiler. Beyond a critical level the ions begin to recombine and the salts settle as a sludge in the bottom of the boiler, or form a hard scale on the tubes.

Distillate. A refined or semi-refined fraction of petroleum obtained in the condensation of a portion of a mixture which has been vaporized by heating. A middle distillate is that portion of petroleum boiling between 330° F (166° C) and about 700° F (371° C.)

Distillation Test: Initial Boiling Point, Final Boiling Point, Total Distillate, Residue, Loss. A test in which the temperature of a sample of liquid fuel is raised gradually until a temperature is reached at which vaporization begins; at this point a certain volume of the fuel has distilled over and both this and the point of final vaporization are noted, i.e. the **initial** and **final boiling points.** As the volume distilled is measured by recondensation, a certain amount of **loss** is inevitable and there will also be some **residue.** The test serves as a guide to fuel volatility at atmospheric pressure. However, under working conditions, pressures in the region of 450 lb/in² are not uncommon; such pressures completely alter the distillation and other characteristics of fuels. Nevertheless, a smooth distillation curve is considered important as it indicates the absence of " heavy ends " in the original fuel.

Distillation Tower. A tall cylindrical fractionating unit used in the first stage of oil-refining operations. A tower contains a number of horizontal perforated bubble cap trays; these trays allow vapour to pass upwards and liquid to flow downwards. Crude oil is first run through coils of pipes lining the walls of a furnace and preheated to about 800° F (427° C); it then passes into the bottom of the distillation tower. At this point all but the heaviest fractions of the crude flash into vapour and pass up the tower. As the various components of the crude have different boiling points and the temperature falls steadily towards the top of the column the rising vapours condense on the trays according to the temperature at which each becomes a liquid. The fractions of gasoline, kerosine, gas oil and diesel oil, lubricating oils and residual fuel oils, are drawn off at different levels and sent to further refining processes which may include further distillation under vacuum.

Distillation Zone. In the combustion of solid fuels, a zone in which fuel is exposed to heat, and the volatiles are distilled out of the solid material. See **Flame Zone; Incandescent Zone.**

District Heating. A scheme in which both heat and hot water are provided from a central boiler plant to an entire housing estate or group of buildings; the consumer enjoys house heating at a comfortable level and a constant supply of hot water. There are two types of district heating scheme: (a) combined electric power and heating plant in which steam is bled from a high-pressure turbine to a water heater; (b) central boiler plant specially designed and constructed for the district heating scheme. It is claimed that consumers enjoy lower running costs per useful therm of heat, although initial capital costs are higher. The scale of the scheme and the density of premises to be served are of great importance to the economics in a particular case. Also known as block heating (a scheme serving one or two blocks of flats or a shopping centre) and as group heating (a scheme serving a group of buildings or small housing estate).

Diurnal. Daily, or recurring every day; the diurnal cycle of air pollution concentrations is of great interest to air pollution authorities.

Diversity Factor. The probability of a number of pieces of equipment being used simultaneously. For example, if 100 electrical appliances of 1 kW each rarely produce a maximum demand in excess of 20 kW, the diversity factor is said to be 1 in 5.

Doctor Test. A method of detecting undesirable sulphur compounds in petroleum distillates, i.e., of determining whether an oil is *sour*, or *sweet*, qq.v. It is a non-quantitative test.

Doctor Treatment. The treatment of gasoline with sodium plumbite solution and sulphur to improve its odour.

Dolomite Brick. A *refractory*, q.v., containing 38 to 42 per cent of magnesium oxide, 38 to 42 per cent of calcium oxide, and 12 to 15 per cent silica.

Dose. The quantity of radiation delivered to a given area or volume, or to the whole body; it is usually measured in rems. See **Roentgen Equivalent, Man.**

Dosimeter. An instrument for measuring the radiation *dose*, q.v.

Double Pole. A description of a switch which is inserted in the wires or "poles" of a circuit. See **Single Pole.**

Dounreay Fast Reactor. A fast nuclear reactor of experimental design located at Dounreay, Caithness, Scotland. The reactor can use *plutonium*, q.v., as a fuel; the coolant is a mixture of liquid sodium and potassium. The reactor is a breeder, that is, it makes more nuclear fuel as it goes along.

Down-draught. A region of severe turbulence formed on the leeward side when wind flows around and over a building. The region of down-draught begins at the top of the windward face of the building, rises to about twice the height of the building, and stretches for about six times the height of the building downwind. Chimney emissions discharged into a down-draught will be brought rapidly to the ground. Chimneys should discharge their gas high enough for them to escape the influence of down-draught. A Committee appointed by the British Electricity Commissioners recommended in 1932 that a power station chimney should be at least two-and-a-half times the height of the tallest adjacent building (usually the power station boiler house), plus an allowance for any difficult topographical features in the vicinity. Today, some chimneys may need to be higher than this to ensure the adequate dispersal of sulphur dioxide. See **Down-wash; Memorandum on Chimney Heights.**

Down-fired Furnace. Or down-shot furnace, a *water-tube boiler*, q.v., in which pulverized coal and primary air is fed vertically downwards from burners near to one side of the furnace roof. The gas outlet is on the other side of the roof and the gases follow a roughly U-shaped path. Secondary air may be introduced concentrically with the primary air jets, or through rows of ports on the side wall adjacent to the burners. This type of furnace is used predominantly with low-volatile coals and anthracite, these being more difficult to ignite than low-rank bituminous coals. Down-fired furnaces may be fired also from opposite ends of the roof, the outlet being at the centre; the result is a W-shaped flame. See Fig. 6.

Down-wash. The drawing down of chimney gases into a system of vortices or eddies which form in the lee of a chimney when a wind is blowing. Down-wash affects the visual appearance of the plume and causes blackening of the stack. In extreme circumstances, it may also assist in bringing flue gases prematurely to ground level. The risks of down-wash may be minimized by: (a) using a round chimney with as small a diameter as possible; (b) avoiding the use

Down-wash

of large overhangs or elaborate ornamentation; (c) discharging the gases from the mouth of the chimney at a sufficiently high velocity. American wind tunnel experiments have indicated that down-wash can be prevented by discharging the gases with a velocity of one and a half times that of the wind passing the top of the chimney. See **Efflux Velocity.**

Dragon High Temperature Reactor. A joint venture between the United Kingdom Atomic Energy Authority and *Euratom*, q.v., a

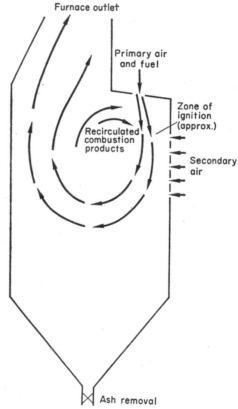

Fig. 6. *General shape of chamber and flow pattern in down-shot water-tube boiler.*

nuclear reactor, q.v., using enriched uranium in ceramic form and graphite as a canning material; temperatures of over 1830° F (1000° C) in the heat cycle are designed for. The reactor is located at Winfrith Heath.

Draught. The difference in pressure which causes air and the products of combustion to flow through the flues of boilers and furnaces; to be effective it must be sufficient to overcome the resistance of the firebed and the friction of the internal surfaces of the system as a whole. It must also be sufficient to achieve the rates of combustion required. See **Balanced Draught; Forced Draught; Induced Draught; Natural Draught.**

Draught Gauge. An instrument for measuring the pressure, negative or positive, in a furnace tube or flue. The simplest form of draught gauge is a glass tube bent to form a U, part filled with water and fitted with a scale; one leg of the U-tube is connected to a suitable point in the boiler flue and the other leg is left open. If the flue is under suction or negative pressure, the pressure of the atmosphere will push the water down in one leg and cause the water to rise in the other; the difference in height of the water in the two legs is a measure of the draught available. Inclined tubes or pointer gauges having magnified scales give a much more accurate reading than the simple U-tube.

Draught Stabilizer. A device to reduce natural draught to that required at the bottom of the flue regardless of the ambient conditions. This is done by allowing air to infiltrate into the chimney through a balanced flap set to the appropriate pressure. The flap is built into a frame the whole of which is inserted into an opening preferably above the flue pipe from the boiler.

Drift Mine. An underground mine in which the coal is reached horizontally from the side of a hill.

"Drift" Theory. A theory as to the mode of origin of coal seams which states that the material of which coal seams are composed drifted there from the areas in which it grew, the site of a seam representing a lake or estuary in which it was deposited. See **"In Situ" Theory.**

Dry. A basis for reporting an analysis of coal; the values for *volatile matter*, *fixed carbon*, and *ash*, qq.v., on this basis are determined by multiplying the values obtained on an *air-dried coal*, q.v., by:

$$\frac{100}{(100 - M)}$$

77

Dry

where *M* is the determined percentage of *inherent moisture* (q.v.). See **Proximate Analysis; Ultimate Analysis.**

Dry, Ash-free (d.a.f.). A basis for reporting an analysis of coal; the values for *volatile matter*, and *fixed carbon*, qq.v., on this basis are obtained by multiplying the values obtained on the *air-dried coal* q.v. by:

$$\frac{100}{[100-(M+A)]}$$

Where *M* and *A* are the determined percentages of *inherent moisture*, and *ash*, qq.v., respectively. See **Proximate Analysis; Ultimate Analysis.**

Dry-back Economic Boiler. See **Economic Boiler.**

Dry-bottom Furnace. A furnace in which the ash deposited is not molten. Only 10 to 30 per cent of the ash leaves through the ash hopper. See **Wet-bottom Furnace.**

Dry-cleaning Process. A process for cleaning small *run-of-mine coal*, q.v., without the use of water. Coal, usually less than 2 in. in size, is passed over a shaking table with a perforated deck through which a current of air flows upwards; the current of air has a velocity sufficiently high to render the coal "fluid", the shale sinking on to the deck while the clean coal rises. The clean coal is then separated from the shale. Coal often comes to the surface too wet to be suitable for treatment by dry methods.

Dry Gas Meter. A gas meter consisting of a rectangular box of tinned steel equipped with two compartments, each containing a bellows, and a system of valves. One valve admits gas to one of the bellows which expands and expels a corresponding amount of gas from the outer compartment. At the end of each movement, the bellows operate a sliding valve so as to reverse the inlet and exit ports; after a bellows has been filled the reversal of the ports allows gas to enter the compartment outside the bellows, thus compressing the bellows which discharges gas to the outlet. The valves are connected to dials indicating the volume of gas passed through. See **Wet Gas Meter.**

Dry, Mineral-matter-free (d.m.m.f.). A basis for reporting an analysis of coal. The value for *volatile matter*, q.v., on this basis is obtained by duly correcting the determined value and then multiplying the result by:

$$\frac{100}{[100-(M+M.M.)]}$$

Where M is the determined percentage of *inherent moisture*, q.v., and *M.M.* the mineral matter in the coal. The value for *M.M.* is obtained by modifying the percentage of ash by the *King-Maries-Crossley Formula*, q.v. See **Proximate Analysis; Ultimate Analysis.**

Dry Quenching. The cooling of hot gas coke by circulating an inert gas through it, the heated gas then being passed through a *waste heat boiler*, q.v. See **Carbonization.**

Dry Saturated Steam. Steam containing neither free moisture nor superheat; this is rarely achieved in practice. See **Dryness Fraction; Wet Saturated Steam.**

Drying. The removal of moisture from a solid by thermal methods in the presence of air.

Drying Plant. Plant designed for the drying of materials; types include: (a) convection driers (drying chambers, rotary, tray, tunnel); (b) contact driers (stationary flat surface, film or drum); (c) spray driers; (d) pneumatic driers; (e) air-swept mills; (f) vacuum driers; (g) radiant and infra-red driers.

Dryness Fraction. The weight of actual steam compared with the total weight of entrained moisture and steam combined. Thus:

$$\text{Dryness fraction} = \frac{\text{Weight of dry steam}}{\text{Total weight of steam and water}}$$

When the dryness fraction is $1 \cdot 0$ then the steam is called *dry saturated steam*, q.v. In practice, dryness fractions may vary from about $0 \cdot 99$ (1 per cent moisture) to about $0 \cdot 90$ (10 per cent moisture); much variation is to be expected.

Dual Firing. Plant with provision for the handling, storing and firing of two kinds of fuel. To equip a power station to fire say oil and coal, increases the capital costs over a station designed to burn oil or coal exclusively. However, the higher cost may be offset by the ability to take advantage of marginal changes in the relative prices of oil and coal, and there is a larger element of security in an electricity supply system capable of flexibility in the event of an unforeseeable reduction in the supplies of one particular fuel. Kingsnorth power station (2000 MW), on the lower Medway, England, is equipped for dual firing by oil and coal.

Duff. *Smalls*, q.v., usually with an upper limit of ⅜ in.

Dulong's Formula. A formula for determining the approximate *calorific value*, q.v., of a fuel by computation from the *ultimate*

Dulong's Formula

analysis, q.v., of the fuel, in British thermal units per pound:

$$Btu/lb = 14{,}450\,C + 61{,}500\left(H_2 - \frac{O_2}{8}\right) + 4000\,S$$

The symbols represent the proportionate parts by weight of the constituents of the fuel (carbon, hydrogen, oxygen and sulphur); the coefficients represent the approximate calorific values of the constituents in Btu/lb. The term $O_2/8$ is a correction applied to the hydrogen in the fuel to allow for the hydrogen already combined with oxygen in the form of moisture. The formula is not considered generally suitable for calculating the calorific values of gaseous fuels.

Dumping Plate. Metal plate which controls the flow of ash over the end of a *chain grate*, q.v., so that it is kept covered. Also called an ash plate.

Dungeness "B" Nuclear Power Station. The Central Electricity Generating Board's advanced gas-cooled reactor (A.G.R.) nuclear power station to be built at Dungeness, Kent, England; it will have a net output of 1200 MW and will be the largest of its kind anywhere in the world. The first in Britain's second nuclear power programme (as outlined in Government White Paper 2335) it is being built by Atomic Power Constructions Ltd. The cost of electricity generated is estimated to be over 25 per cent cheaper than that from the ninth *Magnox*, q.v., power station being built at Wylfa in Anglesey, and over 10 per cent cheaper than that from the most modern coal-fired power station being built at Cottam in Nottinghamshire. The complete reactor, including core, gas circulators and boilers, is housed within a concrete pressure vessel, with the core at the centre, and the circulators under the boilers as shown in Fig. 7. The concrete pressure vessel is cylindrical; an inner pressure cylinder in the form of a bell separates the core from the boilers. The circulators deliver into the region within this cylinder, some of the flow passing downwards through the core to keep the graphite and its restraint structure close to the inlet temperature. The total flow then passes upwards through the fuel channels and charge tubes, and is discharged through ports into the plenum above the pressure cylinder, whence it passes down through the boilers back to the circulator inlets. The internal walls of the concrete pressure vessel are clad with sufficient stainless steel thermal insulation to make it unnecessary to employ a "hot box" as in the Windscale A.G.R. The uranium dioxide fuel is contained in stainless steel cans; these are arranged in

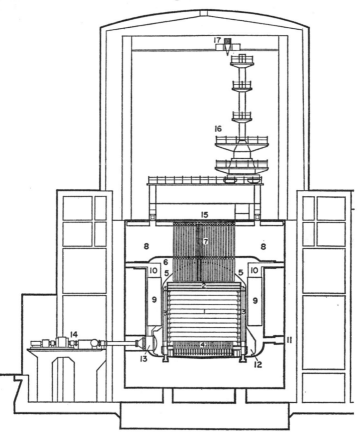

FIG. 7. *Dungeness "B" nuclear power station.*

1. *Core and reflector.*
2. *Top shield.*
3. *Side shield.*
4. *Support structure.*
5. *Pressure cylinder.*
6. *Thermal insulation.*
7. *Fuelling and control standpipes.*
8. *Concrete pressure vessel.*
9. *Boiler.*
10. *Steam outlet.*
11. *Feed water inlet.*
12. *Plenum chamber.*
13. *Circulator.*
14. *Circulator drive.*
15. *Charge face.*
16. *Fuelling machine.*
17. *Charge face crane.*

Dungeness "B" Nuclear Power Station

three-ring clusters of 36 pins within graphite sleeves, to form 40-inch long fuel elements. Two 660 MW turbo-generators operate at the same steam conditions as in modern coal-fired power stations. See **Advanced Gas Cooled Reactor (A.G.R.).**

Net electrical output	1200 MW
Gross generation	1320 MW
Number of reactors	2
Number of turbo-generators	2
Overall station efficiency	41·5%
Type of fuel	36 × 0·57 in. pin clusters
Number of fuel elements in stringer	8
Mean fuel rating for reactor	9·5 MW/tonneU
Initial enrichment	1·47/1·76%
Feed enrichment	1·99/2·42%
Refuelling	continuous on load with axial shuffle
Lattice pitch	15·5 in.
Lattice geometry	square
Active core height	27 ft
Active core diameter	31 ft
Total number of channels	465
Number of fuel channels at equilibrium	412
Channel gas inlet temperature	320° C (608° F)
Channel gas outlet temperature	675° C (1247° F)
Peak can temperature	800° C (1472° F)
Circulator outlet pressure	450 lbf/in²abs.
Number of circulators per reactor	4
Type of circulator	centrifugal
Circulator speed	1500 rev/min
Speed variation	fluid drive coupling
Circulator drive	synchronous motor
Circulator installed power, each	16 500 hp
Type of boiler	once through
Steam pressure	2315 lbf/in²abs.
Steam temperature	565° C (1050° F)
Steam flow	3·7 M lb/h
Reheat pressure	556 lbf/in.²abs.
Reheat temperature	565° C (1050° F)
Condenser vacuum	28·9 in.Hg
Cooling water inlet temperature	14° C (57° F)

Electrical power density-kW/ft³ of
pressure vessel volume   3·0
Electrical power density-kW/ft³ of
building volume   0·07

Durain. Hard and almost lustreless dull coal which presents no pronounced lamination. The two principal components are *micrinite*, and *fusinite*, qq.v.

Dust. Particulate matter of natural or industrial origin which passes a 200 mesh B.S. test sieve (76μ). Dusts which are about 5μ in size or less are respirable and capable of reaching the alveoli of the lungs; such dusts, without necessarily being a nuisance, may constitute a health hazard. See **Grit.**

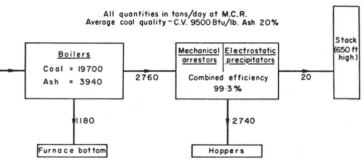

FIG. 8. *Ash flow diagram for a* 2000 *MW power station with a coal consumption of* 19,700 *tons a day at M.R.C.*

Dust Arrester. A device for removing grit and dust from flue gases and other gas streams. In some situations two types of dust arrester may be combined in series. One example is the use of cyclones before bag filters in cleaning the exhaust gases from a copper and alloys foundry. In another example multi-cellular collectors may be used before electrostatic precipitators. This latter approach has been adopted by the *Central Electricity Generating Board*, q.v., in respect of a number of modern power stations, although the trend is now towards using electrostatic precipitators alone. Figure 8 shows diagrammatically the ash flow for a 2000 MW power station using a combination of low efficiency mechanical arresters (about 32 per cent) and high efficiency electrostatic precipitators (about 99 per cent)

83

Dust Arrester

to give a combined dust arresting efficiency of 99·3 per cent. See
**Bag Filter; Cyclone; Dust Arrester Efficiency Test; Dust Burden;
Dust Burden, Measurement of; Electrostatic Precipitator; Fan
Collector; Gravel Bed Filter; High Efficiency Cyclone; Multicellular
Collector; Settlement Chamber; Venturi Scrubber; Wet Washer.**

Dust Arrester Efficiency Test. A test to determine the collecting
efficiency of an arrester designed to remove grit and dust from flue
gas streams. An efficiency test on an arrester plant consists essentially of measuring the amount of dust in the flue gases entering the
plant and the amount in the flue gases leaving the plant over a given
time. The efficiency is then:

$$\frac{\text{Wt. of dust entering} - \text{Wt. of dust leaving}}{\text{Wt. of dust entering}} \times 100\%$$

When testing large plants, the cross-section of the entry duct is
divided into 24 imaginary rectangles. A probe is inserted through
holes provided in the walls of the duct to the centre of each rectangle
in turn and left there for a predetermined length of time. During this
period dust is collected by the probe and deposited in a filter; at the
same time, the gas velocity is also measured. After every rectangle
in the cross-section has been sampled, the total dust collected is
weighed and the total mass flow of gases determined from the velocity
measurements. A combination of this data and the known characteristics of the sampling probe enable the total amount of dust
passing the cross-section during the test to be determined. An
exactly similar procedure is followed at the exit duct from the
arrestor plant; both tests take place simultaneously. The test procedure is generally in accordance with British Standard 893: 1940
*Method of Testing Dust Extraction Plant and the Emission of Solids
from Chimneys of Electric Power Stations.* A typical test on a modern
power station plant requires ten minutes sampling time at each of
the 24 positions or a total of 4 hours sampling. Making allowance
for withdrawing and re-inserting the probe extends this time to 5
or 6 hours. Thus one team can make only one test a day, and a
further one or two days are required to work out and check the
result. The practical difficulties of testing arrester plant also increase
with larger sizes of unit. With 500 MW units it is normal design
practice to divide the gas stream from the boiler into three or four
parallel flows; each parallel flow must be tested independently and
simultaneously. Allowing for supervision, about 20 men would be
required to test the arrester plant of one 500 MW boiler. A 2000 MW

power station would, of course, contain four such units and each may need to be tested under several different operating conditions (e.g. full load, part load, low CO_2, etc.). Simplified methods for the measurement of grit and dust emission are described in B.S. 3405: 1961.

Dust Burden. The weight of dust suspended in a unit of medium, e.g. flue gas. It is customary to express this in grains/ft^3 measured at *normal temperature and pressure* (*N.T.P.*). q.v.

Dust Burden, Measurement of. The determination of the *dust burden*, q.v., in flue gases by appropriate sampling techniques. It is not possible to collect the whole of the material emitted from a stack and it is necessary to obtain a representative sample of it. This implies three stages: (a) the selection of a suitable position from which one or more samples distributed over the cross-sectional area of the flue can be taken; (b) the definition of the technique by which each individual sample is obtained from the flue; (c) the study whereby the results so obtained are converted into reliable estimates of the total emission. Methods of sampling and test procedures are described in B.S. 893 : 1940 and B.S. 3405 : 1961. In respect of dust sampling equipment developed by The British Coal Utilisation Research Association, the sampling nozzle faces " upstream " so that the dusty gases flow into it under the suction of the fan. Rate of flow of gas is metered by a flow measuring device, and controlled by a valve. To ensure a correct sample, it is essential to maintain the same velocity into the nozzle as in the gas stream. This is known as iso-kinetic sampling. It is achieved by first measuring the gas velocity after insertion of the nozzle by measuring the gas flow through the probe. The dust extractor consists of a small cyclone that collects all the grit and dust, and a small filter following it that collects all the fine material. The pressure drop across the cyclone is used as a measure of gas flow. When sampling is completed the dust collected in the extractor is weighed and calculations made, either of total amount of dust emitted, or the concentration of dust in the gas which can be calculated from the weight of the extracted dust and the volume of gas sampled. See **Pitot Tube.**

Dust Deposit Gauge. A widely used instrument for measuring rates of deposition of grit and dust from local sources. It is of considerable value in seeking confirmatory evidence of a nuisance. A British Standard gauge is described in B.S. 1747: Part 1: 1961. It comprises a metal stand, collecting bowl of 12 in. diameter and a 10 litre collecting bottle. The gauge collects only the heavier particles and is

Dust Deposit Gauge

relatively inefficient for small particles. Another type of gauge, known as a " directional deposit gauge ", has been developed by the Central Electricity Research Laboratories in the United Kingdom; it is more efficient as a dust collector and enables a particular source of dust to be more readily identified.

Dust Deposition Rate. The rate of fall-out of grit and dust emitted by a chimney over the surrounding district. A well established formula is that of Bosanquet, Carey and Halton, "Dust Deposition from Chimney Stacks", *Proc. I. Mech.E.* 1950, **162**, pp. 355–366:

$$\text{Average rate of deposition} = \frac{37 \cdot 8 \, Wb}{H^2}$$
$$\times 10^6 \times F\left(\frac{f}{V}, \frac{x}{H}\right) \text{ tons/mile}^2 \text{ month.}$$

where, W = rate of dust emission from stack at M.C.R., lb/s

b = fraction of time during which wind is in direction of 45° sector under consideration

H = height of plume, ft

f = free falling speed of particles, ft/s

V = wind velocity, ft/s

x = distance from stack, ft

The value of $F\left(\frac{f}{V}, \frac{x}{H}\right)$ may be obtained from a graph in the original paper. Dust particles of less than 20μ in size have a negligible free falling speed and are more likely to behave in accordance with the laws of gaseous diffusion; the above formula is considered appropriate only for dusts above this size.

Dust Monitor. An instrument designed to give an indication of the total dust burden in a flue gas, or the size characteristics of a dust. In the A.E.I. Flue Dust Monitor an isokinetically collected sample of gas is drawn through the apparatus; the dust particles are electrically charged, each in proportion to its surface area. These charges are collected and in aggregate form a current proportional to the total surface area of the dust sampled in unit time. The current, after amplification, may be used either to indicate or record, and can operate a warning above a preset level. The instrument is sensitive to fine dust down to less than $0 \cdot 1\mu$.

Dutch Oven. A furnace suitable for burning solid waste chips and sawdust. It consists of a large rectangular chamber, either lined with a refractory material or made of fire-brick, which contains a horizontal grate and a fairly high bridge. The fuel is introduced through one or more holes in the roof, and forms cone-shaped piles on the grate. There are arrangements for the admission of primary, secondary and tertiary air.

Dynamic Viscosity. A measure of resistance to flow; it is the force required to move a plane surface of area 1 cm² over another parallel plane surface 1 cm away at a rate of 1 cm/s, when both surfaces are immersed in a liquid. It is the product of the specific gravity and the *kinematic viscosity*, q.v., in stokes. The unit of dynamic viscosity is the *poise*, q.v.; a smaller unit, the centipoise, is often used where 1 poise = 100 centipoise. See **Viscosity.**

Dyne. The unit of force in the *metric system*, q.v., of units; a force of one dyne, acting on a mass of one gramme, imparts to it an acceleration of 1 cm/s².

E

Ebullition Chamber. A chamber in an *evaporator*, q.v., in which the boiling of water to produce distilled water takes place.

Ecart Probable. Or probable error; one half of the density interval between 25 and 75 per cent recovery as shown in the *partition curve*, q.v.

Economic Boiler. A shell boiler in which the flue gases after passing through the main furnace flues return to the "smoke box" at the boiler front through numerous small diameter fire-tubes, situated above, alongside, or below the main furnace flues. The economic boiler does not require brick flues, as in the *Lancashire boiler*, q.v., although in some cases brick flues have been added to give a third "pass" for the flue gases; it is more common to provide this third "pass" by means of an additional set of fire-tubes within the shell of the boiler. Economic boilers are of wet-back or dry-back design. In the wet-back the rear flue gas reversing chamber is completely surrounded by water; in the dry-back the chamber is set at the extreme rear of the boiler and lined with firebrick. Economic boilers are designed for pressures up to 300 lb/in² and for evaporations of from 700 to 37,000 lb steam/h. Thermal efficiencies are up to 80 per cent, being more efficient than the Lancashire boiler as well as occupying less space. See **Super-economic Boiler.**

Economizer. Or feed-water heater, a device for utilizing the waste-heat from boilers to preheat incoming feed-water thus raising the overall *thermal efficiency*, q.v., of the plant. An economizer comprises banks of tubes placed in the path of the flue gases as they pass to the chimney, after the *superheater*, q.v., but before any *air preheater*, q.v. The external surfaces of the tubes are liable to fouling; in the case of plain tubes scrapers are used which move up and down each tube keeping the surface clean, but in respect of gilled tubes steam or compressed air soot blowers must be used. The temperature of feed-water in an economizer may be raised to within 40° F (22° C) of the water and steam temperature in the boiler.

Eddy Diffusion. The process by which gases, including smoke, diffuse in the atmosphere; molecular diffusion is extremely slow by comparison and may be ignored.

Edeleanu Process. In the petroleum industry, a finishing process based on solvent extraction using liquid sulphur dioxide as a solvent. The process removes undesirable aromatics and other polar compounds such as sulphur, gum and colour constituents; it has been used widely in the manufacture of premium *kerosine*, q.v. The process yields an unpleasant waste *acid sludge*, q.v. The sludge may be decomposed in a special kiln and the *sulphur dioxide*, q.v., converted to sulphuric acid.

Effective Height of Emission. The height above ground level at which a plume is estimated to become approximately horizontal.

Efflux Velocity. The speed at which gases leave a chimney or vent and escape into the general atmosphere; the velocity of discharge should be high enough to avoid any substantial risk of *down-wash*, q.v. Table 5 indicates the minimum chimney velocities recommended for forced and induced draught plant. The most modern of the Central Electricity Generating Board's power stations are designed

TABLE 5—MINIMUM EFFLUX VELOCITIES

Type of draught	Boiler rating, lb/h	Minimum efflux velocity at full load, ft/s
Forced Draught only	—	20
Induced Draught	Up to 30,000	25
Induced Draught	30,000 to 450,000	25/50

Adapted from *Memorandum on Chimney Heights*, British Ministry of Housing and Local Government, 1963 for efflux velocities of 75 ft/s at M.C.R.; this velocity is preserved under part-load conditions by the subdivision of the single chimney commonly adopted into a number of separate flues, each flue serving only one boiler. See **Chimney Heights, Memorandum on.**

Einstein's Equation. An equation which can be used to calculate the energy released in any process, such as *fission*, q.v., in which a loss of mass occurs; it is expressed $E = mc^2$, in which E is energy, m is mass and c the velocity of light.

Electric Arc Furnace. A *steelmaking furnace*, q.v., in which three carbon electrodes are used to carry an electric current from a supply transformer to the steel charge in the furnace bath. Once a circuit is established, the electrodes are withdrawn slightly from the steel so that the current jumps in a lightning flash from the electrode tips to the metal; thus an electric " arc " is struck between metal and electrodes.

Electrical Energy. A capacity for doing work, measured in terms of the *joule*, q.v.; one joule is expended when an electrical power of one *watt*, q.v., is exerted for one second. A 100-W electric light bulb consumes 360,000 J of electrical energy in one hour; this is equivalent to about 341 Btu of heat energy.

Electrical Power. The rate of expenditure of *electrical energy*, q.v.; it is measured in J/s, or watts. Watts = volts × amperes.

Electricity Council. A body set up under the Electricity Act of 1957 to take over the central administration of the electricity supply industry in England and Wales. The Council's statutory duty is " to advise the Minister on questions affecting the electricity supply industry and to promote and assist the maintenance and development by the Electricity Boards of an efficient, co-ordinated and economical system of electricity supply ". It also deals with labour relations, finance and research.

Electrodes. The conductors which convey an electric current into or out of a liquid or a gas.

Electrolytes. Chemical compounds and solutions of chemical compounds, capable of being decomposed by the passage of an electric current.

Electromagnetic Induction. A principle discovered by Michael Faraday in 1831 that if an alternating current, or a direct current of varying strength, is passing through a conductor, then any other approximately parallel conductor in the vicinity will have an electromotive force induced in it. If the ends of the latter are joined so as to form a closed circuit, then an electric current will be induced. In practice, the respective conductors usually take the form of coils of insulated wire. Electric generators, static transformers and induction motors depend upon electromagnetic induction for their operation.

Electromagnetic Pump. A type of pump used for pumping liquid metals; it has no moving parts. The pumping action depends on the interaction between an electric current which flows through the liquid metal and a magnetic field produced over the same region by a magnet. It may be used for pumping a metallic *coolant* in a *nuclear reactor*, qq.v.

Electromotive Force (e.m.f.). A difference in electrical potential which tends to cause an electric current to pass from the point of higher potential to the lower.

Electron. An elementary particle of mass $9·11 \times 10^{-28}$ gramme, carrying a charge of negative electricity of $1·602 \times 10^{-19}$ *coulomb*, which, in the *atom*, moves rapidly in orbit about the *nucleus*, qq.v.

Electron Volt. eV. A unit of energy equal to the energy acquired by an *electron*, q.v., when it is accelerated through a potential difference of one volt. $1 \text{ eV} = 1·602 \times 10^{-19} \text{ J}$. See **Joule.**

Electronic Valve Rectifier. A transformer-rectifier for converting the alternating current of normal electricity supply into high tension direct current electricity; this unit has found wide use in the United States in electrostatic precipitators. See **Electrostatic Precipitator; Rectifier.**

Electroplating. A process by which a metal is deposited, usually as a relatively thin coating, upon the surface of an article which is itself made of a different metal. The main purposes for which electro-deposits are used are to provide: (a) protection against corrosion of the basis metal; (b) a particular surface appearance; (c) good wear resistance. Low voltage direct current from a *rectifier*, q.v., or motor-generator is passed through a solution consisting mainly of a compound of the metal to be deposited. The article to be plated is the "cathode" or negative electrode in the solution; the "anode" or positive electrode is either an insoluble material or more usually the metal to be deposited. Metal is deposited from the solution on to the article.

Electrostatic Filter. An *air filter*, q.v., in which dust particles are given a positive electric charge by an ionizing screen; they are then attracted to the negatively charged filter plates. While offering little air flow resistance the high efficiency of these filters declines as air velocity increases. Some form of after-filtration is usually recommended.

Electrostatic Precipitator. A device for the arrestation and removal of dust from a gas stream. It utilizes the general principle that if a gas is passed between two electrodes one of which is supplied with a very high negative voltage and the other earthed, the gas and any particles of dust in suspension become electrically charged; the dust is attracted to the earthed electrodes where it collects. A mechanical rapping device dislodges the particles and they fall into a collecting hopper beneath the precipitator. The modern precipitator is undoubtedly one of the most efficient means of extracting dust particles from flue gases. Units installed at large new power stations are designed to operate at collecting efficiencies of 99·3 per cent. They are also expensive, at about £0·75 per kW installed. The subdivision of precipitators into zones or sections ensures better performance and flexibility; in large units there may be three parallel banks or more, each bank being divided into three zones in series. Each zone is supplied with separate electrical and rapping equipment. This arrangement allows optimum operating conditions in each zone, and in the event of a failure in one zone ensures that the increase in dust emission will be small. See **Automatic Voltage Control; Deutsch Formula; Dust Arrester; Plate Precipitator; Tubular Precipitator.**

Element. A substance which cannot be decomposed (split up) by chemical changes into simpler substances. Elements are divided into metals and non-metals. Metallic oxides are described as basic and, if soluble in water, as alkalis. Non-metallic oxides are acidic. A compound consists of two or more elements combined together chemically in definite proportions.

Elutriation. The classification or grading of particles effected by movement relative to a rising fluid. A known weight of dust is placed in a receptacle at the base of the apparatus and is then subjected to an upward current of air or water; by varying the velocity of the upward stream it is possible to obtain a series of fractions expressed in terms of the falling velocity of the dust.

Emissivity. The ratio of the rate of loss of heat per unit area of a surface at a given temperature and in certain surroundings, to the rate of loss of heat per unit area of a *black body*, q.v., at the same temperature and in the same surroundings. A black body is a body with an emissivity of unity. The emission of radiant energy from a body depends upon both its temperature and its emissivity. Typical values for emissivity for various substances at 1832° F (1000° C) are: building brick, 0·45; chromium, polished, 0·38; fireclay brick, 0·75; lampblack, 0·96; silica refractory brick, 0·66. See **Absorptivity.**

91

Empirical Formula. The simplest possible formula to describe a chemical compound, an empirical formula indicates only the proportions in which the constituent atoms are present in the molecule. For example, both acetylene C_2H_2 and benzene C_6H_6 have an empirical formula of CH; normal butane C_4H_{10} has an empirical formula of C_2H_5. Thus an empirical formula must be distinguished from a *molecular formula*, q.v., which indicates the actual numbers of the constituent atoms in the molecule. See **Constitutional Formula; Graphical Formula.**

Enamel Firing. A stage in the firing of pottery ware in which the colours or decoration are fired into the glazed ware at a temperature of about 1380° F (750° C). See **Biscuit Firing; Glost Firing.**

Endothermic Gas. A furnace atmosphere and carrier gas used in a wide range of heat treatment processes, including gas carburizing. The use of *liquefied petroleum gases*, q.v., for producing endothermic gas is well established. To produce the gas a mixture of the hydrocarbon feedstock and about one third of the volume of air required for its complete combustion is passed over a nickel catalyst heated to about 1922° F (1050° C). A typical analysis of endothermic gas, each constituent being expressed as a percentage, is: carbon monoxide, 20 to 28; hydrogen, 19 to 41; nitrogen, 38 to 50. See **Exothermic Gas.**

Endothermic Reaction. A chemical reaction accompanied by the absorption of heat. An endothermic reaction occurs during the manufacture of *water gas*, q.v.

Engine Efficiency. The thermal efficiency of an internal combustion engine operating on a constant volume cycle, expressed in the formula:

$$E = 1 - \left(\frac{1}{r}\right)^{y-1}$$

where, E = efficiency
r = compression ratio

$$Y = \frac{\text{Sp. ht. at constant } P}{\text{Sp. ht. at constant } V} \simeq 1\cdot296;$$

P = pressure;
V = volume.

for a weak petrol/air mixture. Thus the higher the *compression ratio*, q.v., the greater the efficiency.

Enthalpy. The total amount of heat that water, steam, air or other gas contains, measured from 32° F (0° C). See **Total Heat.**

Enthalpy-temperature (It) Diagram. A diagram which enables the direct derivation of the initial *enthalpy*, q.v., of any combustion gas to be made, rendering stoichiometric calculations based on the ultimate analyses of fuels unnecessary. Gas temperatures may be directly obtained from the enthalpy of the combustion gases without recourse to the composition or specific heat of the gas.

Entrainment. The collecting and transporting of a substance by the flow of another fluid moving at a high velocity. For example, boiler water may become entrained in the steam leaving the boiler under certain conditions, or dust particles may become entrained in flue bases and carried out of the chimney.

Entropy. In thermodynamics, an index of the availability of heat-energy for producing power. If, in a reversible change, a substance receives or loses a quantity of heat dQ at an absolute temperature T, the substance gains or loses an amount of entropy given by:

$$d\phi = \frac{dQ}{T}$$

where, $d\phi$ = entropy change
dQ = heat added or lost, Btu/lb
T = absolute temperature, °R.

This simple formula covers the case where, for example, water is converted into steam through the addition of *latent heat*, q.v., without change of temperature. If the temperature also varies, from T_1 to T_2, the change of entropy is given by:

$$d\phi = \int_{T_1}^{T_2} \frac{dQ}{T}$$

Equivalent Free-falling Diameter. The diameter of a sphere which has the same density and the same free-falling velocity in any given fluid as the particle under consideration.

Erg. The unit of work or energy in the *metric system*, q.v., of units. It is equal in magnitude to the work done when the point of operation of a force of one *dyne*, q.v., is allowed to move one centimetre in the direction of the force. See **Joule**

Error Curves. A method, developed by Tromp, for assessing coal washing performance. He observed that the shape of the curves derived from *float and sink test*, q.v., data resembled Gaussian error distribution curves. By using these curves he could demonstrate the difference between theoretical and practical results.

Euratom. A European community formed to exploit the peaceful uses of atomic energy. It came into being at the beginning of 1958. Its members are the six members of the European Economic Community (the "Common Market")—Belgium, France, Germany, Italy, Luxembourg and the Netherlands. Euratom coordinates the activities in the nuclear field of its six members by providing common research facilities and a nuclear supply agency, and by establishing basic standards for health protection and safety. Its main institutions are a Commission and a Council of Ministers representing member states. Agreements for cooperation in research and construction have been signed with Britain, the United States and Canada.

European Nuclear Energy Agency (E.N.E.A.). An organization set up in December, 1957, as part of the Organization for European Economic Cooperation, to develop collaboration between the countries of Western Europe in the use of nuclear energy for peaceful purposes. There are eighteen members. The United States, Canada and Japan are associate members. The Agency is concerned with the establishment of a uniform regulating and administrative atomic régime in Europe, especially in relation to health and safety, nuclear liability and the transport of radioactive materials. It also studies the economic aspects of nuclear energy and its place in Europe's overall energy balance sheet. The Agency promotes scientific and technological cooperation between members in the field of nuclear energy. Three major joint undertakings created by the E.N.E.A. are the Eurochemic Company at Mol (Belgium) which reprocesses irradiated fuel; the Halden boiling heavy water reactor project in Norway; and the Dragon high temperature gas-cooled reactor in Britain (at Winfrith).

Eutetic Mixture. That mixture of two or more substances which has the lowest possible freezing point.

Evaporative Capacity. The quantity of steam produced by a boiler expressed in lb/h.

Evaporator. A device for the concentration of solutions or for the preparation of distilled water. Heating steam may pass through a coiled tube surrounded by liquid (weir type) or may surround a bank of tubes through which cold liquid flows (Calandria, Kestner and climbing film types). Evaporators fall into two categories: (a) single effect evaporators, in which steam arising in the evaporation process is simply condensed; (b) multiple effect evaporators, in which steam produced in the initial stage is used as a heating medium in a second evaporating unit, additional "effects" being added until the useful range of heat in the steam is exhausted (hence, double or triple effect evaporators).

Everclean Window. A device developed by the Central Electricity Research Laboratories, Leatherhead, Surrey, England, for keeping the windows of smoke indicating equipment clean. It comprises a honeycomb unit of experimentally determined proportions into which dust will not penetrate. See **Smoke Density Indicator.**

Excess Air. Combustion air supplied in excess of the theoretical or stoichiometric air required for combustion, in order to ensure complete combustion under practical conditions. It is usually expressed as a percentage of the *theoretical air* (q.v.); hence,

$$\text{Excess air, } \% = \frac{(W_a - W_t)\,100}{W_t}$$

where, W_a = amount of air in pounds actually supplied per pound of fuel;

W_t = theoretical or stoichiometric air in pounds

Excess air may be calculated from a knowledge of oxygen content of the waste gases, assuming no combustibles, as follows:

$$\text{Excess air, } \% = \frac{100\,O_2}{21 - O_2}\,K$$

where, $K = 0.96$ for bituminous coal

0.95 for oil

0.90 for natural gas.

Exhauster Fan. Fan used in conjunction with a suction-type pulverized fuel mill to provide carrying air for the pulverized coal.

Exothermic Gas. Gas burned with less than the stoichiometric amount of air to produce an exothermic gas; *town gas*, or *liquefied petroleum gases*, qq.v., may be used for this purpose. Exothermic gas may be "lean" or "rich" according to the greater or smaller amounts of air used respectively. This gas has a wide application in *heat treatment*, q.v., processes, e.g. bright heat treatment of low carbon steels and bright *annealing*, q.v., of copper. See **Endothermic Gas.**

Exothermic Reaction. A chemical reaction accompanied by the evolution of heat. This occurs during combustion processes.

Extraneous Ash. *Ash*, q.v., arising from that part of the *mineral matter*, q.v., associated with, but not inherent in, *coal*, q.v.

F

Fahrenheit Scale. A temperature scale based on three fixed points: (a) the lowest (0°) being the temperature of a mixture of ice, water and sea-salt; (b) the second (32°) being the freezing point of water; (c) the highest (212°) being the boiling point of water at normal atmospheric pressure. The Fahrenheit and Centigrade (Celsius) scales are conventional scales: 32° F = 0° C and 212° F = 100° C, where 1·8 Fahrenheit degrees = 1 Centigrade (Celsius) degree. See **Temperature Scales.**

Fairweather Calorimeter. A continuous recording *gas calorimeter*, q.v.; it is a modified *Boys calorimeter*, q.v., the temperature rise of the water through the calorimeter being recorded electrically in terms of *calorific value*, q.v.

Fan. A pressure-producing device; fans are rated at total water gauge, i.e., actual pressure of "static head", plus pressure equivalent to velocity or "kinetic head". With radial blading, a fan rated at 15 in. w.g. will have 7·5 in. w.g. actual pressure and will deliver the air at a velocity equivalent to a further 7·5 in. w.g. This latter component (H) equals the velocity in ft/s squared, divided by 4440:

$$H \text{ (in. w.g.)} = \frac{v^2}{4440}$$

See **Axial Flow Fan; Backward-curved Fan Blading; Centrifugal Fan; Forced Draught Fan; Forward-curved Fan Blading; Induced Draught Fan; Radial-tip Fan Blading.**

Fan Collector. An *induced draught fan*, and *dust arrester*, qq.v., combined; the dust particles are thrown to the periphery of the stream by the fan and are skimmed off with a proportion of the gas into a *cyclone*, q.v., for separation and collection. The clean gas returning up the centre of the cyclone body to the outlet scroll is led back to the fan inlet.

Fanning. The behaviour of a chimney plume when the air is very stable, the gases quickly reaching their equilibrium level and travelling horizontally with sideways meanderings but with very little dilution in the vertical direction. Fanning produces a thin but concentrated layer of pollution which may impinge on hillsides or tall buildings. See **Fumigation.**

Fanning's Equation. Or D'Arcy formula; a formula for calculating pressure drop due to frictional resistance to flow in pipes:

$$\Delta p = 1 \cdot 295 \times 10^{-3} \frac{flWv^2}{d} \text{ lb/in}^2$$

where, Δp = pressure drop, lb/in^2
f = coefficient of friction
l = length of pipe, ft
W = specific weight of fluid, lb/ft^3
v = velocity, ft/s
d = pipe diameter, in

Farad. A unit of electrical capacitance; it is the capacitance of a capacitor between the plates of which there appears a difference of potential of one volt when it is charged by a quantity of electricity equal to one *coulomb*, q.v.

F.C.C.U. Feed Pretreater. An oil refinery unit for pre-refining the feed stock charged to a *fluid catalytic cracking unit* (*F.C.C.U.*), q.v. The unit usually involves hydrogenation which practically eliminates sulphur and metallic compounds and reduces nitrogen.

Feed Preparation Units. Oil refining units which redistill long residuum from the primary crude distillation units, giving some 50 per cent heavy distillate feed for the catalytic cracker, and 50 per cent of short residuum worked up later for bitumen or blended with other stocks to produce fuel oil. The inlet temperature in the columns approaches 752° F (400° C) and the process is under vacuum. The vacuum is obtained by multi-stage steam ejectors and any gases produced are burnt in plant heaters.

Feed Water. Water suitable for feeding to a boiler.

Feed-water Accumulator. A system in which a supply of boiler feed water is " topped up " by surplus steam not required elsewhere in a plant. By using surplus steam in this way, a reserve of preheated feed water is built-up enabling peak loads of 25 to 30 per cent in excess of average demand to be met.

Feed-water Check Valve. A non-return valve to prevent water from escaping from a boiler should the pressure fall in the feed water supply pipe. Unless a combined stop valve/non-return valve is used, a stop valve must be provided between the non-return valve and the boiler; this enables the non-return valve to be repaired or replaced while the boiler is in operation.

97

Feed-water Heater. A heat exchanger for preheating feed water before it passes to the boiler. See **Economizer.**

Feed-water Meter. A device for measuring the supply of feed water to a boiler; it may consist of a *V-notch meter*, an *inferential meter*, or a *positive displacement meter*, qq.v.

Feed-water Regulator. A feed-water control valve automatically maintaining a constant water level in the boiler drum.

Feedback Control System. See **Closed-loop Control System.**

Feedback Controller. A device which measures the value of a controlled variable in a control system, makes a comparison with a standard representing the desired performance, and as necessary manipulates the controlled system in order to maintain the required relationships. See **Closed-loop Control System; Open-loop Control System.**

Ferric Oxide. Fe_2O_3. An oxide of iron emitted as a dense reddish brown smoke during the "after blow" from a *Bessemer converter*, q.v., and during oxygen lancing in steel-making processes.

Ferricyanide Process. A wet scrubbing process for the removal of *hydrogen sulphide*, q.v., from refinery and petroleum oil gas streams. The scrubbing medium is sodium ferricyanide, $Na_3Fe (CN)_6$, the hydrogen sulphide forming ferrocyanide; the ferrocyanide is oxidized, precipitating sulphur. See **Hydrogen Sulphide Removal.**

Ferrous Metals. Metals in which iron is the main constituent. See **Cast-iron; Steel.**

Fick's Law. A formula giving the molar rate of *mass transfer*, q.v., per unit area of a component A in a mixture of A and B:

$$Na = -D_{AB} \frac{dC_A}{dy}$$

where, Na = molar rate of diffusion per unit area
D_{AB} = diffusivity of A in B
C_A = molar concentration of A
y = distance in direction of diffusion

With turbulent motion, eddy diffusion taking place in addition to molecular diffusion, the formula becomes:

$$Na = -(D_{AB} + E_D) \frac{dC_A}{dy}$$

Film Badge. A piece of wrapped photographic film, often with parts shielded by filters against certain types of radiation, worn by workers liable to be exposed to nuclear radiation. Examination of the film after photographic development enables both the radiation dose and to some extent the type of radiation received by the worker to be determined.

Filter Ratio. The number of cubic feet of gas that will pass through one square foot of filter surface per minute at a given filter resistance; the gas to cloth ratio.

Filter Resistance. The pressure drop across a filtering surface expressed in inches water gauge.

Final Steam Temperature. Steam temperature at the main steam stop valve of a boiler.

Fines. Very fine coal particles, usually less than $\frac{1}{4}$ in.

Fire Point. The temperature of a liquid fuel that will vaporize enough oil to support continuous combustion; it is generally about 20° F (11° C) higher than the *flash point*, q.v.

Firebrick. A *refractory*, q.v., consisting essentially of alumino-silicates and silica, comprising less than 78 per cent of silica and less than 38 per cent of alumina.

Firebridge. A firebrick wall or barrier in a boiler furnace tube to prevent coal from being thrown over the back of the grate into the furnace tube and to prevent air from by-passing the grate at the back. The firebridge is 9 in. thick and the distance from the top of the bridge to the crown of the flue varies from 11 to 15 in. depending on the diameter of the tube.

Firebridge Bearer. Cast-iron support bolted to brackets fixed in the furnace plates of a boiler furnace tube; the end of the bearer is bevelled to suit the firebars.

Firing. Or stoking, the process of feeding *fuel*, q.v., to a furnace or combustion chamber.

Firing by Hand. Methods of hand-firing boilers and furnaces with solid fuel, three of which are in use: *spreader or sprinkler firing*; *coking or dead plate firing*; *alternate side- or wing-firing*, qq.v.

Firing Tools. Tools used by a fireman in tending a furnace; they include (a) shovel, not larger than size 6; (b) poker, or pointed bar; (c) rake, for levelling the fire; (d) hoe, for removing clinker; (e) slice bar, for separating clinker from the grate; (f) pricker bar, for cleaning the spaces between the firebars from below.

Fischer-Tropsch Synthesis. A process in which synthesis gas at about 446 to 626° F (230 to 330° C) and at pressures up to 750 lb/in² is brought into contact with a suitable prepared catalyst (iron oxide, cobalt, nickel or ruthenium) to produce liquid fuels, combustible gas, waxes or organic chemicals. The Fischer-Tropsch process has not been a success economically in most countries. See **Gas Synthesis.**

Fishtail Burner. A gas burner in which gas is emitted from a slot in the blunt end of the tube. Air mixes with the outer fringes of the gas, the resulting combustion heating the gas which has not been exposed to air; the effect of heating is to thermally decompose or crack the gas to carbon and hydrogen. While the hydrogen burns with no visible flame, the carbon particles become incandescent and produce a yellow luminescence. If combustion is incomplete, soot and carbon black are formed. See **Bunsen Burner.**

Fission. The splitting of a *nucleus*, q.v., into two approximately equal fragments. The process is accompanied by the emission of neutrons and the release of energy. While neutron-induced fission is the most common it can also be brought about by bomdardment with heavy particles and by the absorption of photons. It may also occur spontaneously at a very slow rate. See **Neutron; Photon.**

Fission Products. Elements that result from nuclear *fission*, q.v.; in addition to uranium and plutonium these may consist of more than 40 different radioactive elements, e.g. arsenic, barium, cadmium, cerium, iodine, silver and tin.

Fixed Bed Gasifier. A gas generator or producer in which a column of close-packed fuel is gasified.

Fixed Carbon. Carbon which does not pass off in *volatile matter*, q.v., (tarry matter and gases) when coal is heated, remaining in the coke. In a *proximate analysis*, q.v., the percentage of fixed carbon is found by adding the percentages of moisture, ash and volatile matter together and subtracting this total from 100.

Flame. A chemical interaction between gases, accompanied by the evolution of light and heat. See **Cool Flame; Diffusion Flame; Neutral Flame; Sensitive Flame.**

Flame Impingement. The impingement of a flame or atomized oil against the side of a combustion chamber; this may lead to the formation of carbon deposits and smoke.

Flame Speed. The speed of propagation of a flame under certain prescribed conditions. If the flame speed is higher than the velocity of the gas at the burner, back-firing may occur; if it is slower the flame will be extinguished.

Flame Stability. The degree to which a flame maintains its correct position in relation to the burner; if a flame does not maintain its correct position it is said to be unstable.

Flame Temperature. A property of *flame*, q.v., which depends on the calorific value of the gases, the latent heat in steam, the heat losses due to radiation and the dissociation of gaseous molecules, and the volume and specific heat of the total gaseous products. It may be represented:

$$Tf = \frac{Q}{V}$$

where, $Tf =$ flame temperature

$Q =$ net calorific value of gases $+$ sensible heat of gas and air $-$ heat lost by radiation and dissociation

$V =$ volume of products $\times$ sp. ht. at constant pressure.

Flame Zone. In the combustion of solid fuels, a zone in which volatile matter is burning after being ignited in the *incandescent zone*, q.v. See **Distillation Zone.**

Flash Point. The lowest temperature at which a product gives off just sufficient vapour to form an inflammable mixture with air under the conditions of a standard test. It is usually a statutory requirement that flash point shall not be less than 150° F (66° C) in respect of various types of fuel oil; in the case of inflammable products with flash points of less than 150° F (66° C) special regulations relating to safety in storage and handling must be complied with. A flash point may be " open " or " closed " depending upon whether the test apparatus is used with a cover or not; the " open " flash point is several degrees higher than the " closed ". See **Abel Flash Point Apparatus; Open Flash Point; Pensky-Martens Flash Point Apparatus.**

Flash Steam. Steam produced from water when the pressure is suddenly reduced.

Flasks. Fifty-ton containers for carrying irradiated nuclear fuel elements; each flask contains some 2 to 2½ tons of used fuel elements which are transported to chemical separation plant.

Flat Rate Tariff. A *tariff*, q.v., for the supply of electricity consisting solely of a single unit charge; used chiefly by very small users of electricity for lighting and power.

Float and Sink Test. A test for determining the possibility of efficiently cleaning a coal by a gravity separation process; a prepared sample of coal is suspended in a series of liquids of increasing density from 1·3 to 1·8 by increments of 0·1. The percentages, and the ash contents, of the floats are determined at each stage, the results being recorded graphically as " washability curves ". See **Error Curves.**

Floating Roof. A special roof which floats upon the liquid in a storage tank, thus eliminating vapour space irrespective of the level of the liquid. If the advantages of floating roofs are to be fully realized there must be effective sealing between the tank wall and the roof. Seals may consist of polyurethane foam, synthetic rubber tubing filled with paraffin or light fuel oil, or be of a mechanical nature comprising perhaps a ring of steel shoes pressed firmly against the shell by springs. Floating roofs are generally considered not necessary for kerosine, gas oil, fuel oil and lubricating oils. They are necessary, however, for all fractions lighter than kerosine and for crude oil; it is from these materials that a significant contribution to oil refinery air pollution would otherwise arise.

Flowmeter. A measuring instrument utilizing an orifice plate, *venturi tube*, q.v., or flow nozzle, to show the rate of flow in a pipeline.

Fluid Catalytic Cracking Unit (FCCU). An oil refinery unit which enables more gasoline to be obtained from the crude oil than is "naturally" present. In this unit vaporized oil, air and a catalyst of clay-like material (alumina-silica-microspheroids) are circulated at high temperatures through a complex system of pipes and chambers. Cracking takes place in a reactor; the cracked gases pass to a fractionating tower for distillation and are fractionated into gas, gasoline, and light and heavy cracked gas oils. The catalyst passes from the reactor to a regenerator where air is used to burn off carbonaceous matter deposited on the catalyst; once the carbon is burned off, the catalyst is re-activated and may be used repeatedly. Catalytic cracking has made it possible to produce more than twice as much gasoline from a barrel of crude as can be made by simple distillation. See **Thermal Cracker.**

Fluid-Coke Process. An oil refinery process in which hot residual oil is sprayed on to externally-heated seed coke in a fluid bed; the fluid coke is removed as small particles. See **Delayed-coking Process; Petroleum Coke.**

Fluidity. The reciprocal of viscosity. In the *metric system*, q.v., the unit is known as the "rhe"; it equals 1/poise.

Fluidized Bed. A bed of solid particles through which air or gas is blown upwards so that the bed assumes the properties of a fluid.

Fluidized Gasification of Coal. See **Winkler System.**

Fluorescein. A liquid dye suitable, for example, for detecting condenser tube leaks.

102

Fluorescent Lamp. A tubular *discharge lamp*, q.v., internally coated with a powder which fluoresces under the action of the discharge producing a shadowless white or coloured light.

Fluorine. F. An *element*, q.v., present in rocks and soils and hence in fuels, fluxes and raw materials used in industry; at. no. 9, at. wt. 18·9984. Fluorides occur in British coals in concentrations ranging up to 175 ppm; the coals containing the higher proportion of fluorine are found in Kent, Staffordshire and South Wales. Brickworks and other branches of the ceramic industry emit fluorides. It is also emitted in steelmaking in which fluorspar is used as a flux. Fluorine is an extremely active element; the use of the word "fluorine" usually means some compound of this element.

Fly Ash. Non-combustible particles suspended in flue gas. See **Pulverized Fuel Ash.**

Fog. The condensation of water in the atmosphere and consequent formation of water droplets. By international agreement fog is defined as visibility below 1100 yards. In general it does not become a major hindrance or nuisance to the public until the visibility falls below 220 yards, i.e. thick fog. Fog is an essential ingredient of *smog*, q.v., as originally defined.

Foot-candle. The unit of intensity of illumination; it is the illumination produced by one candle power at a distance of one foot.

Forced Draught. The supply of combustion air to a furnace at a pressure greater than that of the atmosphere, utilizing a fan. The forced draught pressure required for stokers ranges from 1 to 4 in. w.g. See **Draught.**

Forced Draught Fan. A *fan*, q.v., through which air is drawn before entering the furnace. Air may be delivered below the furnace grate, or above and below simultaneously.

Forward-curved Fan Blading. A design of blading for centrifugal fans which is considered one of the most suitable for forced draught fans; its point of maximum efficiency is at the load that it is designed to carry, and it has no reserve of pressure with which to overcome dirty boiler conditions. Initial cost is lower but running costs are slightly higher compared with *backward-curved fan blading*, q.v. See Fig. 9. See **Fan.**

Four-wire Distribution. A system of distribution employed on 3-phase a.c. systems; it consists of three "phase" wires and one neutral wire.

Fractional Distillation. The separation of the components of a liquid mixture by vaporizing the mixture and collecting the fractions

Fractional Distillation

FIG. 9. *Forward-curved fan blading*.

which condense in different temperature ranges. See **Distillation Tower.**

Free-burning Coal. *Coal*, q.v., which does not cake in the fuel bed. See **Caking Coal.**

"Free-carbon". A loose description for microscopic particles of resinous material present in creosote-pitch. At normal storage temperatures of 80 to 90° F (27 to 32° C) these remain dispersed; if overheating occurs in certain conditions the particles tend to form a deposit. See **Coal Tar Fuels.**

Free-falling Velocity. The rate of fall of a particle through a still fluid.

Free Moisture. Moisture in a solid fuel which can be removed by evaporation or by centrifugal force. Free moisture is acquired when coal is washed, and is that moisture which remains after normal draining. Free moisture within certain limits improves the fuel bed by lowering its resistance to the passage of air and stimulates combustion. See **Inherent Moisture.**

Free-swelling Index. A measure of the behaviour of coal when heated rapidly; it may be used as an indication of the coking characteristic of a coal when burned as a fuel.

Freons. A range of chemicals which have proved particularly useful as refrigerants because of their non-explosive characteristics; they consist mainly of fluorine (and other halogen) derivatives of hydrocarbons.

Frequency. The cycles produced per second by an alternating current electricity supply. See **Alternating Current.**

Friability Test. A test to measure the tendency of a coal to break during handling; it is determined by the use of a standard tumbler.

"From and at 212° F." A basis for comparing the evaporative capacities of boilers. The total heat in steam depends upon both the pressure and temperature of the steam and boilers are assumed for comparative purposes to generate steam at normal atmospheric pressure (14·7 lb/in² abs.), i.e., from water at 212° F to steam at the same temperature. In respect of any boiler, an "equivalent evaporation" may be calculated as follows:

$$E = \frac{W(T - t)}{L}$$

where, E = equivalent evaporation "f. and a. 212° F"
W = weight of water evaporated, lb
T = total heat in steam per pound, Btu
t = heat in feed water per pound, from 32° F (0° C), Btu
L = latent heat of steam at 212° F (100° C) and normal atmospheric pressure.

Froth Flotation. A wet cleaning process for separating "dirt" in the form of sandstone, shale, clay, pyrites, calcite and so on from *run-of-mine coal*, q.v. In this process, fine coal and water are beaten up by an impeller with a "frothing agent" such as creosote. Clean coal floats to the surface while dirt sinks to the bottom of the flotation cell. The froth is separated from the fine coal in a vacuum filter.

Fuel. A substance used to produce heat, light or power, usually by *combustion*, q.v., in air or oxygen. Most natural or primary fuels such as coal, wood, peat, oil and natural gas are made up of compounds of carbon, hydrogen and oxygen, in association generally with mineral ash, moisture, sulphur and nitrogen. *Nuclear fuel*, q.v. stands in sharp contrast by virtue of its nature and mode of heat release. See **Fuels, Primary and Secondary.**

Fuel/Air Flow Ratio. A relationship which has been employed as a means of combustion control in gas and liquid fuel fired steam boilers.

Fuel Cell. An electrochemical cell which operates by utilizing the energy of a spontaneous chemical reaction, e.g., the combustion of a carbonaceous, hydrogen or hydrocarbon fuel by oxygen from the air. The reactants are fed to the cell at rates proportional to the amount of electrical energy which is required. The electrodes used in fuel cells are generally gas-adsorption electrodes. See **Cell.**

Fuel Economics. The study and analysis of the whole pattern of forces determining the supply of, and demand for, fuel and energy of all kinds.

Fuel Economy. The theory and practice of efficient fuel utilization.

Fuel Efficiency. The proportion of the potential heat of a *fuel*, q.v., which is converted into a useful form of energy.

Fuel Element. An assembly consisting of fissile material (for example, natural or enriched uranium) contained in a can; the latter often has fins to improve the heat transfer between the element and *coolant*, q.v.

Fuel Injection Equipment. In respect of the *diesel engine*, q.v., equipment to inject fuel oil into the cylinders; injection begins during the compression stroke (at 10 to 20° of crank angle before top dead centre) and continues during the power stroke. The equipment controls the quantity of fuel injected into each cylinder and ensures that the fuel is distributed in a suitably atomized form. Formerly achieved by using compressed air, fuel injection is today described as "solid injection". Injection equipment falls into two main categories, the *Common Rail System*, and the *Jerk System*, qq.v.

Fuel Ratio. The ratio of *fixed carbon*, to *volatile matter*, qq.v.; a ratio sometimes used in coal classification systems.

Fuels, Primary and Secondary. Primary fuels are those forms of energy obtained directly from nature, e.g. coal, oil and natural gas. Secondary fuels are those derived from primary fuels, e.g. coke and town gas.

Fuidge Diagram. A diagram defining the combustion characteristics of rich gases; increasing air/gas ratios are plotted against thermal inputs.

Fulham-Simon-Carves Process. A process for removing sulphur dioxide from flue gases using ammonia liquor as the washing medium. The process consists essentially in scrubbing the gases with ammonia liquor to produce ammonium salts which by autoclaving are converted into ammonium sulphate and sulphur. The advantage of the process is that it produces saleable products. See **Battersea Gas Washing Process; Howden-I.C.I. Process; Reinluft Process.**

Fume. Airborne solid particles arising from the condensation of vapours or from chemical reactions; fume particles are generally less than 5μ in size, respirable, and visible as a cloud. They may be emitted in the following processes: (a) volatilization; (b) sublimation; (c) distillation; (d) calcination, and (e) chemical reaction.

Fumigation. A rapid increase in *air pollution*, q.v., at ground level caused by turbulence of the atmosphere created by a rising morning sun following a nocturnal *inversion*, q.v., in which pollutants have become concentrated aloft; very high ground level concentrations may be experienced for an hour or more, sometimes at a distance of many miles from the source of the pollution. This fumigation effect during the break-up of an inversion is due to the restoration of turbulence initially at ground level which gradually penetrates into the stable layers above which are still acting as a "lid" inhibiting the upward dispersal of pollutants. Fumigation was originally described by E. W. Hewson in an article "The meteorological control of atmospheric pollution by heavy industry", *Quart. J. Royal Met. Soc.* **71**, 266–282, 1945. See **Fanning.**

Furnace. An enclosed space in which heat is produced from the chemical oxidation of a *fuel*, q.v., or from another source of energy.

Furnace Ash. A mixture of *ash*, *riddlings*, and *clinker*, from the *combustion*, qq.v., of coal or coke which collects in the furnace bottom, as distinct from *fly ash*, q.v., which is carried forward through the flues.

Furnace Oil. The heaviest grades of natural or cracked petroleum oils.

Furnace Rating. The maximum heat input of a furnace expressed usually as Btu/h ft^3 furnace volume.

Fusain. A dull, very friable, charcoal-like coal *lithotype*, q.v., of silky texture, occasionally fibrous; an important constituent of coal. Known also as "mineral charcoal".

Fuse. A safety device consisting of a few inches of relatively fine wire, mounted in a suitable holder or contained in a cartridge and connected as part of an electrical circuit. If the current exceeds a predetermined value, the fuse wire melts (i.e., the fuse "blows") and thus obviates damage to the circuit protected by the fuse.

Fusible Link. A piece of low melting-point metal forming part of a wire connected to a shut-off valve in an oil supply line. A safety device, if a fire occurs and the link melts the valve automatically shuts.

Fusible Plug. A plug designed to give warning of overheating due to insufficient water in a boiler. The plug consists of an outer body of bronze or gunmetal, with a central conical passage of up to $\frac{3}{8}$ in. diameter. The passage is closed with a core secured by an annular lining of fusible alloy so that the plug may drop clear if the lining melts. The fusible metal should melt at a temperature of not less than 150° F (66° C) in excess of the saturated steam temperature at the design pressure of the boiler. If the water level falls causing undue heating, the metal melts, the core falls out and water and steam escape into the furnace. B.S. 759:1967 describes fusible plugs and specifies the positions in which they should be placed.

Fusing Temperature. In relation to ash, the temperature at which it softens. Ashes that fuse in the range 1900° to 2200° F (1038 to 1204° C) are designated "low fusing"; those fusing in the range 2200° to 2600° F (1204 to 1427° C) as "medium fusing"; and those above 2600° F (1427° C) as "high fusing". In general, ash with a low softening temperature is likely to form *clinker*, q.v.

Fusinite. A powdery, black mineral-charcoal; a subordinate component of *clarain*, q.v., in which it may occur as irregular bands or as disseminations in *micrinite*, q.v., but a principal component of *durain*, q.v.

"Fyrite" CO_2 Apparatus. A portable instrument for measuring the *carbon dioxide*, q.v., content of flue gases. A known sample of gas is introduced into the apparatus by means of a rubber bulb. When the instrument is turned upside down, the sample is bubbled through a solution of potassium hydroxide; a final reading for CO_2 being obtained in only one minute. Accuracy is within $\frac{1}{2}$ per cent of CO_2, the scale being graduated 0 to 20 per cent.

G

Galloway Boiler. An early type of boiler similar in outward appearances to the *Lancashire*, q.v.; however, after the firebridge the two furnace flues combined into a single flue which contained water tubes set across the flue at various angles throughout the remaining length of the boiler. It was claimed that these tubes, up to thirty in number, promoted a better circulation of water while increasing heating surface and steaming power. Also known as a "breeches-flued" boiler.

Gamma Rays. Electromagnetic radiation emitted by a *nucleus*, q.v.; similar to X-rays but usually of shorter wavelength. Gamma rays are exceedingly penetrating.

Gas Act, 1965. An Act giving the *Gas Council*, q.v., a monopoly to distribute gas in Britain; thus all North Sea natural gas produced must be offered to the Council.

Gas Burner. See **Aerated Burner; Bunsen Burner; Diffusion Flame Burner; Direct Burner; Fishtale Burner; Nozzle-mix Burner; Post-aerated Burner; Pre-aerated Burner; Premix Burner; Tunnel Mixing Burner.**

Gas Calorimeter. An instrument for the determination of the *calorific value*, q.v., of fuel gases. See **Bomb Calorimeter; Boys Calorimeter; Fairweather Calorimeter; Junkers Calorimeter; Sigma Calorimeter.**

Gas Coal. *Bituminous coal*, q.v., possessing a relatively high *volatile matter*, q.v., and suitable for the manufacture of *coal gas*, and *coke*, qq.v.

Gas Coke. Coke produced in the manufacture of *town gas*, by the *carbonization*, qq.v., of coal.

Gas Council. A central body set up under a British Act of 1949. The Council comprises the chairmen of the twelve regional gas boards which manufacture and supply town gas to industrial and domestic consumers. The Gas Council is responsible for finance and general policy in respect of the industry.

Gas Engine. An internal combustion engine in which a gaseous fuel is mixed with air to form a combustible mixture in the cylinder and fired by spark ignition. See **Heat Engine.**

Gas Flow Formula. A formula for calculating the flow of gas through a gas burner orifice:

$$Q = 1655 \; KA \; \sqrt{\frac{P}{S}}$$

where, Q = gas flow, ft³/h at N.T.P.
$\quad K$ = coefficient of discharge of the orifice (0·61 to 0·9)
$\quad A$ = area of the orifice, in²
$\quad P$ = pressure of the gas, in. w.g.
$\quad S$ = specific gravity of the gas (air = 1).

Gas Governor. A device for controlling gas pressures or volumes in mains or appliances. The purpose of a constant pressure governor is to maintain a constant pressure at the outlet which is independent of pressure fluctuations at the inlet. Simple pressure governors are of a diaphragm or bell type. A constant volume governor maintains the same gas rate irrespective of small pressure fluctuations.

Gas Laws. Laws indicating the relationships between pressure, temperature and volume, while two of those quantities are changing. See **Boyle's Law; Charles's Law.**

Gas Meter. See **Dry Gas Meter; Wet Gas Meter.**

Gas Oil. A term designating the heaviest vaporizable portion of petroleum, ordinarily boiling from about 550° F (288° C) to 1000° F (538° C) or more; a distillate with a viscosity and boiling range between kerosine and lubricating oil. So called because of its extensive use in the gas industry to produce *carburetted water gas*, q.v. Its use has been extended to gas turbines in the electricity industry and for furnace use generally. A typical *Central Electricity Generating Board*, q.v., specification for gas oil is: specific gravity at 60° F (15° C), 0·85; viscosity Redwood No. 1 at 100° F (38° C), 32 to 40 s; closed flash point, 150° F (66° C); gross calorific value, 19,200 Btu/lb (10,665 Chu/lb); sulphur, 0·4 to 1·0 per cent; pour point, 10 to 20° F (−12·2 to −6·65° C).

Gas Producer. Plant for the production of *producer gas*, q.v.; it usually comprises a vertical cylindrical water-cooled shell fitted with a charging hopper at the top and a fire grate at the bottom containing tuyeres for the admission of an air/steam blast. The plant may be followed by cleaning equipment for the removal of dust and tar. Fuels suitable for gas producers include gas coke and oven coke, anthracite and dry steam coals, and non-caking or weakly caking bituminous coals. The reactions in the fuel bed of a producer are shown in Fig. 10. Steam is added to the air blast to assist in reducing clinker formation and to promote the formation of carbon monoxide and hydrogen. In the secondary distillation and reduction zone a reaction known as the *water gas shift reaction*, q.v., takes place whereby carbon monoxide is replaced by its own volume of hydrogen.

Gas Purification. The removal of unwanted or injurious components of a gas, particularly *hydrogen sulphide*, q.v.

Gas Recycle Hydrogenator (GRH) Process. A *Gas Council*, q.v., pressure process for manufacturing rich gas from *naphtha*, q.v.; the gas may be supplied direct to the consumer after blending.

Gas Reduction Process. A process in which a gas (e.g. carbon monoxide, hydrogen or methane) is employed to reduce ores to metals.

Gas Separation Units. In oil refining, units which receive cracked gases from the catalytic cracker and by a sequence of absorption and distillation under pressure, yield streams of propane-propylene, butane-butylene and light gas. These streams are treated in a Girbotol or similar plant to remove from them hydrogen sulphide and other undesirable sulphur compounds. These gases may be liquefied by compression and are known as *liquefied petroleum gases*, q.v. They are stored in spherical tanks at refineries. See **Girbotol Process.**

Gas Synthesis. The catalytic reaction of a mixture of hydrogen and carbon monoxide (synthesis gas) to produce liquid fuels, combustible gas, waxes or organic chemicals. The synthesis gas may be made by

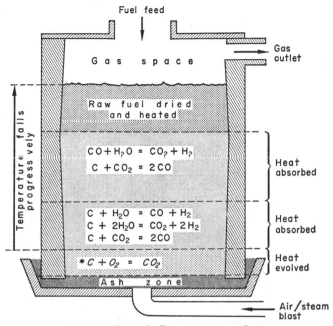

FIG. 10. *Thermal effects in a gas producer.*

* Reaction shown in italics is exothermic. All other reactions are endothermic.

reforming methane or by the gasification of coal with steam. See **Fischer-Tropsch Synthesis.**

Gas Turbine. A *heat engine*, q.v., working on the principle of compression and expansion of a gas, normally air. The essential difference between a gas turbine and other forms of heat engine is that the compression and expansion take place across rotating parts, not by reciprocating motion as in a *diesel engine*, q.v. The process is generally continuous and this characteristic gives a very even torque for power transmission. The basic gas turbine consists of a compressor, a combustion chamber for heating the compressed air and a turbine. In the "open cycle" turbine the air, after compression and expansion, exhausts to atmosphere. In the "closed cycle" turbine the working fluid, generally air, is continuously recirculated, the heating necessary for expansion being effected by an air heater. The gas turbine is capable of running on a wide variety of fuels—gases, liquids and solids. See **Steam Turbine; Turbo-jet; Turbo-prop.**

Gasification. The conversion of the combustible material in a solid or liquid fuel into a combustible gas.

Gasifier. Any unit of equipment in which the process of *gasification*, q.v., is carried out. See **Lurgi Gasifier.**

Gasoline. Or petrol; a complex mixture of light hydrocarbon blending stocks, the characteristics of which are varied to suit all motorcars and aircraft in all seasons and climates. During refining, motor gasoline distills within the temperature range of 100 to 400° F (37·8 to 204·4° C); aviation gasolines are blends of high-octane number stocks and distill within the temperature range of 100 to 325° F (37·8 to 162·8° C). Four grades supply the needs of virtually all conventional aircraft, the grades being based on octane number requirements. Fuels for jet turbine engines boil in the 100 to 600° F (37·8 to 315·6° C) range and have no octane number requirement.

Gasoline Additives. Small amounts of compounds of substances added to gasoline to reduce the tendency to knock, i.e. to improve the *octane number*, q.v. The compounds react with the activated oxygenated intermediate compounds whose decomposition products result in knocking. Generally, the most effective additive has been found to be tetraethyl lead (TEL); about 2 ml/gal are added to gasoline in the United Kingdom, and up to 4 ml/gal in America and Europe. TEL is used with a scavenger to prevent the formation of deposits; ethylene dichloride and dibromide are used and these form volatile lead halides. Another additive, tetramethyl lead (TML) is now used in some super grade gasolines. A Joint Report by the U.S. Public Health Service and the Petroleum Industry in 1965 concluded that lead in gasoline resulting in atmospheric lead contamination was not a hazard to health in the three cities studied— Cincinnati, Philadelphia and Los Angeles. Other additives include organic phosphates to prevent spark plug fouling; phenols and amines as anti-oxidants to reduce gum formation; and unsaturated organic acids and orthophosphates as anti-rust agents.

Gasoline Engine. An internal combustion engine in which a gasoline-air mixture, provided by the *carburettor*, q.v., is drawn by vacuum through an intake manifold to the combustion chamber of each cylinder; it is then compressed and ignited by an electric spark. Each piston in a four-stroke engine moves up and down twice to produce a power impulse:

(a) Intake stroke—the piston descends drawing into the cylinder a gasoline/air mixture produced by the carburettor

(b) Compression stroke; the piston ascends compressing the gas to about one-sixth or one-eighth of its original volume in the cylinder

(c) Power stroke—following ignition by an electric spark the expansion of the gases forces the piston down;

(d) Exhaust stroke—the piston ascends and the exhaust gases are expelled to atmosphere via exhaust manifold, muffler and tail pipe. The thermal efficiency of a petrol engine varies between 18 to 25 per cent. See **Diesel Engine; Motor Vehicle Exhaust Gases.**

Gauge Pressure. Pressure above atmospheric pressure as indicated on a *pressure gauge*, q.v.

Geiger-Müller Counter. A device for counting the number of charged particles or photons by the ionization which they produce in a gas between two electrodes. The operating voltage is sufficiently high for the primary ionization to cause breakdown of the gas resulting in the production of a relatively large output pulse. See **Photon; Radiation Detector.**

GeV. Giga electron volt $= 10^9 eV$.

Girbotol Process. A wet scrubbing process for the removal of *hydrogen sulphide*, q.v., from refinery and petroleum oil gas streams. The scrubbing medium is an aqueous solution of ethanolamines, usually diethanolamine, the reaction being:

$$(CH_3CH_2OH)_2 NH + H_2S \rightarrow [(CH_3CH_2OH)_2 NH_2]HS$$

Absorption is carried out in packed towers, and regeneration in a bubble-cap tower using stripping stream. See **Hydrogen Sulphide Removal.**

Gland Steam. Steam used to prevent air from entering a *turbine cylinder*, q.v., between the turbine shaft and the casing.

Glass Wool. An insulating material, but not suitable for surfaces above 930° F (499° C); it is available in mattress form or in rigid semicircular sections.

" Gloco ". A *gas coke*, q.v., meeting British Standard specification 3142 as a fuel suitable for coke-burning domestic open-fires.

Glost Firing. A stage in the firing of pottery ware in which the glaze is fired on to the biscuit ware at a temperature of about 1960° F (1070° C). See **Enamel Firing; Biscuit Firing.**

Grab Sample. A sample of gas or liquid taken over a very short period of time, a time insignificant compared with the total duration of the operation.

Graded Coals. Coals classified into size groups by screening. See Table 6.

TABLE 6—GRADED COAL SIZES

Grade	Typical screen size in.	Possible size range in.
Large cobbles	6 × 3	8 to 3
Cobbles	4 × 2	5 to 2
Trebles	3 × 2	$3\frac{1}{2}$ to $1\frac{1}{2}$
Doubles	2 × 1	$2\frac{1}{4}$ to 1
Singles	1 × $\frac{1}{2}$	$1\frac{1}{2}$ to $\frac{1}{2}$
Peas	$\frac{1}{2}$ × $\frac{1}{4}$	$\frac{3}{4}$ to $\frac{1}{4}$
Grains	$\frac{1}{4}$ × $\frac{1}{8}$	$\frac{7}{16}$ to $\frac{1}{8}$

Gradient of Potential Temperature. See **Potential Temperature, Gradient of.**

Graphical Formula. A chemical formula which indicates the position of every atom and linkage in the molecule. A simplified form of graphical formula for normal butane is shown in Fig. 11. See **Constitutional Formula; Empirical Formula; Molecular Formula.**

Graphite. A dense, rigid, allotropic form of carbon used as a *moderator*, in a *nuclear reactor*, qq.v.

Graphite Sleeve. In respect of a *nuclear reactor*, q.v., a hollow graphite cylinder used to contain and support the fuel rods in the channels of the reactor.

Grate. A device to support the fuel bed, allow sufficient *primary air*, q.v., to pass through it with as even a distribution as possible, and assist in the separation of ash from the burning fuel. Grates may be horizontal or inclined, stationary or movable, and operated manually or automatically. With stationary grates the ash and clinker has to be removed with hand tools; a rocking bar grate discharges the ash through the bars into the ashpit by mechanical action; with a self-cleaning grate the fuel moves to the rear, the ash and clinker being discharged over the back end.

Gravel Bed Filter. A *dust arrester*, q.v., consisting of filter beds of one or more layers of abrasion-resisting material such as gravel; the filtering material effectively removes dust from the gas stream passing through it. The captured dust is removed from the filter bed

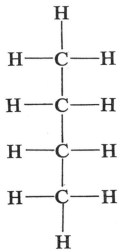

Fig. 11. *Graphical formula for normal butane.*

by a vibrating system which shakes the spring-supported filter bed containers. A gravel bed filter may be used in continuous operation at temperatures up to 650° F (343° C); the pressure drop is low.

Gravity Feed System. A system for supplying oil to an *oil burner*, q.v.; oil is pumped to an overhead tank from which it flows by gravity to each burner. As the oil is not under high pressure, the system is suitable only for blast atomizers. See **Blast Atomizer.**

Gray-King Assay. A test for determining the caking and swelling properties of coal. As described in B.S. 1016, 20 grammes of coal are heated in a silica tube in a standard furnace at a defined rate of temperature increase to 1110° F (600° C). The appearance of the resulting residue is then compared with a series of standard cokes to which the letters A to G are allocated; the series ranges from non-caking up to highly-caking coal. Coals of type A are non-caking; those which give a hard strong coke, but have not swollen are described as type G. For coals more strongly swelling than those of type G subscripts are added, G_1, G_2, and upwards. The subscripts are determined by the addition of inert material to the coal.

Grey Cast-iron. A *cast-iron*, q.v., in which the carbon is in a free state as graphite flakes; it is easily machined.

Grid. A high-voltage 132 kV (132,000 V) electricity transmission system operated by the *Central Electricity Generating Board*, q.v. It is called "the grid" because its originators likened the proposed transmission system to a grid-iron. Its function is to provide a link between every power station in the country, thus ensuring supplies, achieving greater economy in generation and keeping spare plant to a minimum. See **Super-grid.**

Grindability Index. The relative ease of pulverizing a coal. See **Ball-mill Method; Hardgrove Machine Method.**

Grit. Particulate matter of natural or industrial origin retained on a 200 mesh B.S. test sieve (76μ); such particles are visible when deposited and likely to cause eye irritation. The particles are non-respirable and do not penetrate to the depths of the lungs; they can be, however, a source of nuisance. See **Dust.**

Grit Carry Forward. The amount of grit and dust passing through a system or process suspended in a medium, e.g. flue gas. For example, in *Gas Purification Processes*, by Nonhebel, Newnes, 1964, Dransfield and Lowe give the following figures for the carry forward of *fly ash*, q.v., in the gas stream from various types of furnace, expressed as a percentage of the ash in the original fuel: Cyclone furnaces, 15 to 20; slag tap furnaces, 45 to 55; dry-bottom pulverized fuel furnaces, 80 to 85.

Grizzly. Screening equipment used in the surface preparation of coal. It consists of a number of sloped parallel bars. The separating size is determined by the width of the openings between the bars, the length of the bars and the slope. This type of screen is used at mine-preparation plants for the production of modified *run-of-mine coal*, q.v., or for the removal of undersize coal ahead of a crusher. Also known as a gravity bar screen.

Gross Calorific Value. The number of British Thermal Units (or Centigrade heat units) liberated when one pound of coal or oil, or a cubic foot of gas, is completely burned in oxygen saturated with water vapour, the final products being carbon dioxide, sulphur dioxide, nitrogen and **liquid** water. In the condensation of the water vapour all *latent heat*, q.v., is recovered. The gross calorific value is usually determined in a *bomb calorimeter*, or a *gas calorimeter*, qq.v. The heating value of a unit weight of oil fuel may be determined by the U.S. Bureau of Mines equation:

$$\text{Gross c.v.} = 22{,}370 - 3789\ d^2\ \text{Btu/lb}$$

where d is the specific gravity of the oil, $60°$ F/$60°$ F. See **Calorific Value; Net Calorific Value.**

Group Heating. See **District Heating.**

Guillotine Damper. An adjustable plate normally installed vertically in the flue between the furnace and the stack, and counterbalanced for easier operation. It may be operated manually or automatically. See **Damper.**

Guillotine Door. A door which is fitted to the hopper outlet of a *chain-grate stoker*, q.v., which controls the thickness of the fuel bed on the grate. Also called an "outlet gate".

Gulf HDS Process. A process developed for the hydro-desulphurization of residual stocks in oil refineries.

Gum. In the petroleum industry, a rosin-like insoluble deposit formed during the deterioration of petroleum and its products, particularly gasoline.

Gun-type Burner. A *pressure jet burner*, q.v., suitable for central heating boilers; it is fully automatic operating under the command of a thermostat. Ignition is by an electric spark.

Gunmetal. An alloy of copper and tin; lead and nickel are often added. It is used where resistance to corrosion or wear is required, e.g., as in gears, bearings and steam pipe fittings.

Gustiness. A form of turbulence set up near the ground by obstacles presented to the direct flow of air by the surface and its irregularities.

H

Hafnium, Hf. An *element* q.v.; at. no. 72, at wt. 178.49, with a high neutron-capture cross-section. It is found in zirconium ores and must be removed before *zirconium*, q.v., is used as a canning material for the fuel elements in a *nuclear reactor* q.v.

Haldane Apparatus. Fuel gas analysis apparatus in which the gas is confined in semi-capillary tubes over mercury and the volumes before and after each absorption are read through a travelling lens to ±0.0001 ml.

Half-life. The time taken for one-half of the atoms of a radio-active isotope, q.v., to disintegrate. Each isotope has a unique half-life which lies between less than 10^{-6} s and more than 10^6 years according to the isotope. Iodine 131 has a half-life of 8 days; caesium 137 a half-life of about 30 years; radium a half-life of 1580 years. The half-life is related to a disintegration constant (λ), as follows: $t_{\frac{1}{2}} = 0.693/\lambda$. The disintegration constant is the fraction of the number of atoms of a particular radioactive isotope which disintegrate in unit time.

Hall Oil Gasification Process. A cyclic and non-catalytic process for the manufacture of a rich gas from oil. The plant comprises two generators containing *chequer-brickwork*, q.v., connected to each other. During the blow stage, air is admitted to one generator in which it burns off the carbon deposited during the previous " make " before passing through the system restoring temperatures; the waste gas passes through a *waste-heat boiler*, q.v., and then to the chimney. Steam is then admitted to the system in the same direction, initially for purging purposes and then to mix with an incoming oil spray; the cracking process is completed as this steam/oil vapour mixture passes down through the chequer-brickwork of the second generator. After the oil is shut off the system is again purged with steam. The whole process is then repeated in the reverse direction. The constituents of a typical gas, expressed as percentages, are: hydrocarbons, 32; methane, 28; hydrogen, 20; carbon dioxide, 5; nitrogen, 15. The calorific value is over 1000 Btu/ft³ (555 Chu/ft³).

Halo Method. A geochemical method used in the exploration for oil. Soil surveys are conducted by measuring the hydrocarbon content and mineralization in subsoil samples; eight samples per square mile are preferred. Significantly high values are plotted; if they form a pattern such as an aureole or halo, the area is considered positive, the petroleum deposits being roughly outlined by the pattern.

Hand Picking. The cleaning of *run-of-mine coal*, q.v., by hand; large screened coal exceeding 4 in. in size is placed on a slowly moving " picking belt ", while workers pick out obvious pieces of shale or dirt.

Hard Asphalt (or Asphaltenes) Test. A chemical test to determine the quantity of asphaltic material in a liquid fuel. The test is made by determining the percentage of material which cannot be dissolved by a certain reagent, usually a petroleum ether of specified characteristics. The quantity of asphaltenes indicated varies with the type of reagent. In practice, high quality distillate fuels have a negligible asphaltene content. Fuels containing asphalt are liable to form carbon deposits in the cylinders of high speed engines, causing deterioration of exhaust valves, piston ring sticking and liner wear. Slow running engines are less prone to this problem, due to the longer period available for combustion.

Hard Coal. *Coal*, q.v., relatively resistant to degradation on handling. See **Soft Coal.**

Hard Coke. Metallurgical coke produced in a *coke oven* q.v., to meet the requirements of the iron and steel industry.

Hardgrove Machine Method. A method for determining the *grindability index*, q.v., of a coal; the Hardgrove Machine is a grinding mill of ring-roll design. The index is based on a standard coal that is assumed to have an index of 100.

Hardness. A characteristic of water representing the total concentration of calcium and magnesium ions. Hardness is expressed fundamentally in terms of the chemical equivalents of metal ions capable of precipitating soap; it is also expressed in terms of the equivalent amount of calcium carbonate.

Hartridge Smokemeter. An instrument developed to give an instantaneous direct reading of smoke density on a 0 to 100 scale, specially designed for use with diesel and other internal combustion engines. The principle of the meter is based on a comparison of the density of a column of smoke with a column of clean air. Figure 12 shows the smoke tube and the clean air tube which are optically and dimensionally identical. At one end of the tubes is a 12 V, 48 W light source, and at the other end a photoelectric cell; these are so mounted that they can be moved together from the smoke tube to the clean air tube by a control lever. The photocell is connected to a micro-ammeter with a smoke density scale 0 to 100, representing the percentage of light absorbed. A sampling pipe connects the instrument to the vehicle exhaust. The smoke density scale is set to zero for the clean air tube; the control lever is then moved to the smoke tube to give an immediate reading.

Heart Cut. A narrow range cut usually taken near the middle portion of the oil being distilled or treated. See **Cuts.**

Heat Balance. A means of determining the thermal efficiency of a process and indicating the origin and magnitude of heat losses. It takes account of the heat supplied to the system in the heating value of the fuel, and in the preheating of fuel, combustion air, feedwater,

Heat Balance

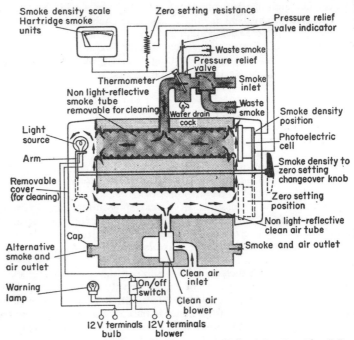

FIG. 12. *Diagram illustrating the working principle of the Hartridge-B.P. smokemeter.*

furnace or charge; and of the heat output by way of absorption in water, steam, or charge, radiation losses, normal stack losses, losses due to inefficient combustion, loss due to combustible matter in the ash, and other losses not readily accounted for. See **Sankey Diagram.**

Heat Drop. The difference between the total heat in the inlet steam and the total heat in the outlet steam of an engine or turbine. In an ideal engine, which would incur no losses, the energy that should be available is called the *adiabatic heat drop*, q.v.

Heat Engine. A machine for converting heat energy into mechanical energy. See **Diesel Engine; Gas Engine; Gasoline Engine; Gas Turbine; Steam Engine; Steam Turbine.**

Heat Exchanger. A device for arranging the transfer of heat from a hot substance to a cooler substance, while keeping the two substances separate. They fall into two groups: (a) those in which the heat transfer process is continuous within the same heat exchanger, e.g. recuperators, condensers, air preheaters and economizers; (b) those in which the transfer is intermittent within the same heat exchanger, e.g. regenerators and blast-furnace stoves, the heat storing solid being alternately heated by the hot fluid and cooled by the fluid to be heated. See **Air Preheater; Blast-furnace Stove; Cascade Heat Exchanger; Condenser; Economizer; Howden-Ljungstrom Air Preheater; Pebble Stove; Recuperative Air Preheater; Regenerator; Superheater.**

Heat Meter, BCURA. A meter for the measurement of heat consumption by the occupants of individual flats in a block, or of houses in a district heating scheme, under development by the British Coal Utilisation Research Association. The basic principle of the meter is that a small fixed proportion (say 1/200) of the "return" water is diverted through a side arm, and reheated, by means of a small electric heater and control system, to the temperature of the water at entry to the dwelling. The proportionate flow is achieved by means of two orifices in parallel, a small one in the side arm, and a larger one, across which the side arm is connected, in the main return pipe. An electricity meter measures the energy input to the small heater and this reading can be directly converted to heat consumption.

Heat Pump. A device for transferring heat from a low temperature medium to a higher temperature medium. For example, heat may be extracted from flowing water by a refrigerant which is then compressed, transferring its heat to another flow of water; the refrigerant passes through an expander to begin the process afresh.

Heat Rate. In respect of a power plant, the heat consumption per kWh of current produced:

$$\text{Heat rate (Btu, Chu/kWh)} = \frac{\text{Heat supplied in fuel, Btu (Chu)}}{\text{Energy generated, kWh}}$$

In a large conventional power station with a thermal efficiency of 37·5 per cent, the heat rate is approximately 9100 Btu (5055 Chu). One kWh = 3412 Btu (1896 Chu).

Heat Release Rate. The amount of heat liberated in a *combustion chamber*, q.v., usually expressed as Btu/h ft^3.

Heat Retaining Arches. Arches of refractories often built over grates at the front, intended to radiate absorbed heat on to incoming green fuel to aid *ignition*, and *combustion* qq.v.

Heat Transfer. See **Conduction; Convection; Inverse Square Law; Radiation; Stefan-Boltzmann Law.**

Heat Treatment. The thermal treatment of metals and alloys after production to improve their physical properties, e.g. the treatment of steel by normalizing, hardening and tempering, or the *annealing*, q.v., of metal after cold working. Heat treatment may involve the chemical treatment of surface layers, e.g. *carburizing*, and *nitriding*, qq.v. Furnace atmospheres often need careful control to prevent surface attack by way of oxidation, scaling, *decarburizing*, q.v., or sulphur penetration.

Heavy Ends. The highest boiling portion present in the distillation of petroleum.

Heavy Water. D_2O. Water in which the molecule contains two atoms of *deuterium*, q.v. It is present in ordinary water to the extent of about 1 part in 42,000,000. Substantially pure heavy water is used as a *moderator*, in some types of *nuclear reactor*, qq.v.

Helium. He. An *element*, q.v.; at. no. 2, at wt. 4·0026; it is the lightest inert gas. It has a very low neutron-capture cross-section and good heat transfer properties and has been considered as a *coolant*, q.v., for nuclear reactors.

Henry. A unit of electrical inductance; it is the inductance of a closed circuit in which an electromotive force of 1 *volt*, q.v., is produced when the electric current in the circuit varies uniformly at a rate of 1 *ampere*, q.v., per second.

Hertz. Unit of frequency; the number of repetitions of a regular occurrence in one second.

Heterogeneous Reactor. A *nuclear reactor*, q.v., in which the fissile material and moderator are arranged as discrete bodies, usually in a regular pattern. The fissile material is contained in the fuel elements. See **Fuel Element.**

High Efficiency Cyclone. A development of the simple *cyclone*, q.v.; the inlet takes the form of a true volute, as distinct from the tangential entry in the simple cyclone, thus reducing the disturbance to the gases already rotating within the body. The rotational movement within the cyclone is also prolonged by a lengthened body and cone. This type of cyclone may involve a pressure drop of up to 4 in. w.g., but particles down to about 20μ can be collected with a high degree of efficiency. See **Dust Arrester.**

High Energy Fuels. Special fuels with performance characteristics superior to those of hydrocarbon fuels; designed for the propulsion of space rockets and missiles. Such fuels include hydrogen, beryllium, boron, diborane, pentaborane, decaborane and alkylborane with calorific values ranging from 61,500 to 25,000 Btu/lb.

High-heat Value (HHV). Synonymous with *gross calorific value*, q.v.; a term most often used in the United States of America.

High Speed Mill. A mill for the production of *pulverized fuel*, q.v.; coal is ground by attrition and impact applied by hammers or pegs rotating inside a casing at speeds ranging from 1000 to 4000 rev/min. This type of mill is most suitable where the output required is less than 1 ton/h.

High Steam and High-low Water Alarm. A safety mounting, often incorporated in a lever type *safety valve*, q.v., designed to give audible warning if the water level in a boiler becomes either dangerously high or low. The two main classes of alarm are: (a) internal for shell boilers; (b) external, for water-tube boilers. An internal alarm consists of two floats, one at each end of a long pivoted arm suspended from the crown of the boiler shell. The low water float lies on the surface of the water at normal level, the high water float being suspended clear of the surface. If the water level falls, the low level float drops; the pivoted arm moves and operates a steam whistle. If the water level rises to an excessive degree the top float becomes buoyant moving the pivoted arm in the same direction as before, the alarm again sounding. External alarms may be of the float or thermostatic types. In the float type a separate chamber is mounted at normal working level and connected to the steam and water space. A float in the chamber responds to changes in the level of the water in the boiler, operating a steam whistle at predetermined high and low positions. In the thermostatic type, two rods are utilized, one of these being normally immersed in the steam and the other in the water. Should the water level rise covering the upper rod, the rod contracts operating an electric alarm; if the water level falls, immersing the lower rod in steam, this rod expands and similarly operates the alarm.

High Temperature Deposits. Deposits on boiler steam tubes and superheater tubes consisting of re-fused ash, alkali-matrix deposits and phosphate deposits. These deposits are associated with high fuel bed temperatures. Stoker fired boilers, with high fuel bed temperatures, are much more liable to high temperature deposits than are pulverized fuel boilers which operate with lower combustion temperatures and a high *grit carry forward*, q.v. See **Low Temperature Deposits.**

High Temperature Plastic. An insulating cement made of mineral-wool fibres processed into nodules and then dry-mixed with clay. It is mixed with water for application and sets as a tough fibrous insulation. It is claimed that the temperature limit for this material is 1800° F (982° C).

High Voltage. Means a voltage normally exceeding 650 volts.

Hilt's Law. In respect of coal deposits a statement of fairly general application that where there has been relatively little earth movement, coal seams having remained more or less undisturbed, the oldest and lowest seams contain less volatile matter than the relatively younger seams above; regular progression being the general rule. Valid as a generalization, important exceptions have been noted.

Hoffmann Kiln. A continuous ring tunnel *brick kiln*, q.v., of annular longitudinal arch design, the kiln has the appearance of an endless tunnel, access to which is gained through a dozen or more doorways or "wickets". The latter are assumed to divide the kiln into a number of "chambers" in which the bricks to be fired are set. Fuel is fed through holes in the roof or arch of the kiln; there may be 16 to 20 fireholes to each chamber. The kiln may be coal or oil-fired. A kiln may measure from 200 ft to over 400 ft in length. Firing proceeds round the kiln so that the bricks in each chamber are subjected to three stages: (a) preheating; (b) firing; (c) cooling.

"Homefire". A six-sided fluidized char binderless briquette, some three inches in diameter and a quarter of an inch thick; a reactive solid fuel burning with a bright flame, and suitable for open domestic fires. It is produced by the *National Coal Board*, q.v., at Coventry, England. See **Authorized Fuels; Smoke Control Area.**

Homogeneous Reactor. A *nuclear reactor*, q.v., in which the fissile material and moderator are uniformly mixed in the solid state or dispersed in a solution or slurry.

Horizontal Boiler. A shell boiler, which may be equipped with simple internal furnace tubes as in the examples of the Cornish and Lancashire boilers, or be equipped with multi-fire-tube systems as in the examples of the *Economic*, *Package*, and *Marine boiler*, qq.v. A horizontal boiler may or may not require a brick flue setting.

Horizontal Retort. A closed chamber used for the manufacture of *town gas*, by the *carbonization*, qq.v., of coal. The retort is oval or D-shaped, constructed in silica brick or fireclay, about 20 ft long, 24 in. wide and 16 in. high, with metal doors at either end. Often grouped in double columns of five retorts, each group being known as a "bed"; a series of beds forms a retort house "bench". Each retort holds about 17 cwt of small coal. The retort is heated by *producer gas* q.v., generated from coke in a step grate producer serving each bed. A heating cycle takes about 12 hours, a temperature of about 1800° F (1000° C) being attained.

Horsepower. An engineering unit of power, equal to a rate of working of 33,000 foot pounds per minute. Thus an engine of one horsepower carries out an amount of work equal to $33,000 \times 60 = 1980 \times 10^3$ ft lb/h. The term horsepower is still occasionally applied to boilers, although an inappropriate term in this context; the amount of steam required by a steam engine for one indicated horsepower may vary from 15 to 50 lb steam/h. The size of a boiler should be linked therefore with the amount of steam required per hour and not with the horse-power required from the engine. Where the term is still applied to boilers, one horse-power is often defined as 34·5 lb steam/h "*from and at* 212° *F*", q.v.

Hot Face Insulation. The insulation of the internal surfaces of a furnace or hot gas duct; it became possible through the development of bricks which combine high refractoriness with good insulation properties.

Hot Gas Efficiency. In respect of a *gas producer*, q.v., thermal efficiency calculated as follows:

$$\text{Hot gas efficiency, } \% = \frac{\text{Total heat of gas}}{\text{Total heat of fuel}} \times 100$$

where total heat means the sum of potential heat (calorific value) plus sensible heat due to preheating. Hot gas efficiencies are of the order of 90 per cent. See **Cold Gas Efficiency.**

Hot Rolled Steel. Steel that is passed, while red hot, through a rolling mill.

Hot Soak. That portion of the *gasoline*, remaining in the *carburettor*, of a *gasoline engine*, q.v., which, after operation, is evaporated by the heat remaining in the stationary engine.

Hot Well. A sump tank to which hot *condensate*, q.v., may be returned and from which the boiler feed pump draws supplies.

Howden-I.C.I. Process. A process for removing sulphur dioxide from flue gases. In operation at Fulham power station in London, before the Second World War, it consisted of a closed circulation system using water with considerable additions of chalk. The process removed over 97 per cent of the sulphur dioxide in the flue gases, converting it to calcium sulphite which oxidized to calcium sulphate. The sulphur extracted was not recoverable in useful form and several hundred tons of sludge were produced every day. The process has not been resumed. See **Battersea Gas Washing Process; Fulham-Simon-Carves Process; Reinluft Process.**

Howden-Ljungstrom Air Preheater. A regenerative *air preheater*, q.v., consisting of a revolving rotor containing heating surfaces. The flue gases are directed axially through one side of the rotor, while air for combustion passes through the other side, the rotor revolving slowly; the heat transferred to the rotor heating surfaces by the flue gases is absorbed by the incoming air.

Humic Coal. *Coal*, q.v., derived from plant debris; includes most *common banded coal*, q.v.

Humidity. See **Absolute Humidity; Relative Humidity.**

Hydrazine. N_2H_4. An oxygen removing agent suitable for treating boiler feed water; it does not add to the *dissolved solids*, q.v. See **Sodium Sulphite.**

Hydrocarbons. Organic compounds consisting of carbon and hydrogen only. They are subdivided into aliphatic and cyclic hydrocarbons according to the arrangement of the carbon atoms in the molecule. The aliphatic hydrocarbons are in turn subdivided into: (a) paraffins; (b) olefins; (c) diolefins, etc., according to the number of double bonds in the molecule. The cyclic hydrocarbons are subdivided into: (a) aromatics; (b) naphthenes or cyclo-paraffins. In all types of hydrocarbon, hydrogen atoms may be replaced by other atoms making the formation of a virtually endless number of compounds possible. See **Aromatic or Benzene Series; Naphthene or Cyclo-paraffin Series; Olefin Series; Paraffin Series.**

Hydrocracking Unit. A process for cracking heavy hydrocarbons to light products in the presence of a high partial pressure of hydrogen and a special catalyst. This process can convert gas oil completely into gasoline and lighter fractions, or convert gas oils into high grade middle distillates. All hydrocracked products are virtually sulphur free.

Hydro-desulphurization. A process used to refine cracked gas oil from a catalytic cracker. It consists in heating the oil and allowing it to trickle through beds of catalyst in an atmosphere of hydrogen. Reactions take place and the hydrogen sulphide formed is separated from the refined oil and delivered to the sulphur recovery plant. The desulphurized oil is blended with straight run gas oils from the crude distillation units to yield the automotive diesel fuels used by tractors, buses and lorries.

Hydrodynamic Lubrication. Full film or fluid lubrication, a state of ideal lubrication which exists when two surfaces in relative motion are completely separated by a fluid viscous film of oil; the film is induced and sustained by the relative motion of the surfaces. See **Lubrication.**

Hydroelectric Power. The use of water power to drive water turbines which in turn drive electricity generators; the degree of utilization of water power varies from nil in some countries to over 50 per cent in Switzerland. See **Pumped Storage.**

Hydroforming. An oil refinery unit which converts or "reforms" low octane heavy naphtha from the crude distillation process into motor spirit of higher quality. This operation is carried out in an atmosphere of hydrogen at high pressure utilizing a fluidized catalyst; the catalyst is regenerated after use.

Hydrogasification. The *hydrogenation*, q.v., of coal or oil to produce mainly gaseous hydrocarbons.

Hydrogen. H. An *element*, q.v.; at. no. 1, at. wt. 1·00797; it combines with oxygen to give the reaction:

$$2H_2 + O_2 \rightarrow 2H_2O$$

Hydrogen has a *calorific value*, q.v., of 61,500 Btu/lb but only about 342 Btu/ft^3 at N.T.P.

Hydrogen or H$_2$ Treater. A hydrogenating plant, often used in oil refineries to desulphurize or otherwise purify hydrocarbons.

Hydrogen Sulphide. H$_2$S. A colourless gas with a density of 1·19 in relation to air and a characteristic foul odour of rotten eggs. It arises in the decomposition of organic material. Other sources include oil refineries, sulphur recovery plants, some metallurgical processes, and various chemical industries using sulphur-containing compounds. Hydrogen sulphide is New Zealand's principal air pollution problem both from indigenous sources and from organic wastes associated with the primary industries (timber pulping, meat packaging, skin curing, etc.). It is emitted from the ground in the Rotorua area of the North Island and is present in the mineral waters of that area. Hydrogen sulphide is irritating to the eyes and respiratory tract; it leads ultimately to death through paralysis of the respiratory centre of the brain. Though the odour of hydrogen sulphide is readily recognizable in low concentrations, the detection of dangerous concentrations by smell is unreliable and unsafe as olfactory fatigue occurs quickly at high concentrations. The gas is corrosive to many metals and even when present in the atmosphere in concentrations below the level of physiological significance it discolours lead paints. Hydrogen sulphide tends to be a localized problem; in ordinary combustion processes the gas is readily burned to suphur dioxide. Significant concentrations of hydrogen sulphide, expressed in parts per million by volume, are: odour threshold, 0·13 to 1·0; powerful sickening smell like rotten eggs, but harmless, 2 to 3; maximum allowable concentration in an atmosphere in which a man works for 8 hours at a time, 20; headache, giddiness, sickness, loss of energy, disturbance of vision, above 50; maximum allowable concentration for 1 hour, 170 to 300; loss of consciousness and death in a short time may occur in some people, 500; a few breaths will kill at once by action of gas on central nervous system, 1000.

Hydrogen Sulphide Removal. In respect of *town gas*, the removal of *hydrogen sulphide*, qq.v., by reaction with moist ferric hydroxide in the form known as "bog iron ore". The ore is placed in trays in shallow purification boxes or in towers. The reaction is:

$$2Fe(OH)_3 + 3H_2S \rightarrow 2FeS + S + 6H_2O$$

The hydrogen sulphide concentration in the gas should be reduced to 1 ppm by volume. See **Claus Kiln; Ferricyanide Process; Girbotol Process; Manchester Process; Seeboard Process; Shell Phosphate Process; Thylox Process.**

Hydrogenation. A chemical reaction with *hydrogen*, q.v., to produce a product containing an increased proportion of hydrogen. See **Coal Hydrogenation.**

Hydrolysis. The decomposition of compounds by interaction with water.

Hydrometer. An instrument for the measurement of *specific gravity*, q.v. or density. A hydrometer may be graduated in degrees, each degree indicating half an ounce of solid matter dissolved per gallon; or graduated to show parts of solid matter dissolved in 10,000 parts of water or more. The instrument is therefore of direct use in measuring the density of boiler water which changes with the total amount of dissolved solids in it.

Hydroxylation. The formation of hydroxylated compounds by the reaction of hydrocarbons and oxygen, as in a *premix burner*, q.v., or a bunsen burner; these compounds become aldehydes which during combustion become CO_2 and H_2O.

Hygroscopic. Readily absorbing and retaining moisture.

I

I.C.I. Steam Naphtha Reforming Process. An oil gasification process using the light distillate, naphtha; the process has two basic stages: (a) sulphur removal; (b) primary reforming with steam at pressure. A hydro-desulphurization process first reduces the more complex sulphur compounds to hydrogen sulphide. The sulphur is then removed by zinc-oxide based absorbents. The reforming part of the process takes place in a tubular furnace in the presence of superheated steam and nickel catalyst. The reaction is continuous and no periodic regeneration is required. All steam requirements are derived from waste heat recovery systems.

Ignafluid Grate. A grate, similar to a *chain grate*, q.v., over which coal fines may be efficiently burned in suspension. The coal enters the furnace through a port in the front wall; a pulsating flow of air from the underside of the grate holds the fuel bed in suspension. Ash is deposited on the combustion faces of the furnace, fuses and flows down on to the grate where it is carried to the back of the furnace and deposited as clinker. The *grit carry forward*, q.v., is high; efficient mechanical centicell grit arresters are required to collect the grit which is then blown back into the furnace. It is claimed that the ignafluid grate will burn efficiently fine coal with an ash content of up to 45 per cent.

Ignition. The beginning of combustion; ignition is effected by raising the temperature of a fuel to a point at which the rate of burning provides the heat essential for the process to continue. Burning does not occur below this temperature which varies with different fuels. Some examples of ignition temperatures are: anthracite, 930° F (500° C); bituminous coal 750 to 800° F (400 to 425° C); hard coke, 930 to 1200° F (500 to 650° C); gas oil, 638° F (336° C); carbon monoxide, 1191 to 1216° F (644 to 658° C); hydrogen, 1080 to 1095° F (580 to 590° C).

Ignition Delay. See Ignition Lag.

Ignition Lag. Or ignition delay; the period between the injection of fuel into a *diesel engine*, q.v., cylinder and its subsequent ignition. Fuels which have a short ignition lag are described as "high ignition quality" fuels. Engine factors such as compression ratio, fuel and air inlet temperatures, fuel injection time, air turbulence and engine speed also influence ignition lag.

Ignition Line. A characteristic feature of the fuel bed on a *chain-grate stoker*, q.v., a distinct line across the surface of the fuel bed representing the division between the distillation stage and the combustion stage of the fuel. See **Ignition Plane.**

Ignition Plane. A characteristic feature of the fuel bed on a *chain-grate stoker*, q.v., a plane along the grate towards the rear starting at the *ignition line*, q.v., caused by the ignition penetrating into the fuel bed. After the ignition plane has reached the grate, the coke burns on the after-part of the grate with a non-luminous flame.

Immature Coal. A general term for low rank *coal*, q.v., See **Mature Coal.**

Impulse-type Turbine. A turbine machine in which steam is expressed in fixed blades or nozzles, the change of direction giving an impulse to the moving blades.

"In Situ" Theory. A theory as to the mode of origin of coal seams which states that a coal seam occupies more or less the same site on which grew the original plants from which it was derived. See **"Drift" Theory.**

Incandescent Zone. In the combustion of solid fuels, a zone in which the non-volatile fixed carbon is above ignition temperature. See **Distillation Zone; Flame Zone.**

Incinerator. Equipment in which solid, semi-solid, liquid or gaseous combustible wastes are ignited and burned. An incinerator may consist of one, two or more refractory lined chambers, interconnected by flues. The purpose of an incinerator is to reduce the bulk of the waste products; after incineration the ash residue may be buried, or carried out to sea and dumped.

Indirect-fired Combustion Equipment. Plant in which the flame and all products of combustion are separated from direct contact with the material being processed by means of metallic or refractory walls, as in heat exchangers, melting pots and muffle furnaces. See **Direct-fired Combustion Equipment.**

Induced Draught. The extraction of flue gases from a furnace by means of a fan situated between the back end of a plant and its chimney; *induced draught fans*, q.v., are used when the natural draught produced by a chimney is not sufficient to draw in all the air required for combustion, or insufficient to remove the products of combustion as quickly as they are formed. See **Draught.**

Induced Draught Fan. A *fan*, q.v., placed near the base of a chimney to draw air through a furnace and remove the products of combustion. As all the hot products pass through the fan, about one and a half times the volume of gases has to be handled compared with an equivalent *forced draught fan*, q.v. In addition to requiring more power, induced draught fans have to withstand the comparatively high temperature of waste gases, some of which are corrosive.

Induction Heating. The production of heat utilizing the principle that an eddy current will be induced in an electric conductor which is subject to a changing magnetic field. The magnetic field is produced by a coil carrying an *alternating current*, q.v., the work to be heated being placed within the coil. The efficiency of conversion of electrical energy into heat by this method rarely exceeds 50 per cent, but the method offers many advantages. Frequencies used for induction heating range from fifty up to a million cycles per second.

Inertia. The reluctance of a body to change its state of rest or of uniform velocity in a straight line. Inertia is measured by mass when linear velocities and accelerations are considered; and by moment of inertia for angular motions, i.e. rotations about an axis.

Inferential Meter. A self-contained meter for measuring feedwater flow on the delivery side of a feed pump; the meter may be of the disc or rotary type. See **Feed-water Meter.**

Inflammability, Limits of. In respect of gases, the lowest and highest ratios of fuel/air mixture at which *ignition*, q.v., can take place. Thus a coal gas/air mixture will not ignite if the gas is less than 5·3 per cent by volume, or higher than 31 per cent by volume, of the total mixture. These are known as the "lower" and "higher" limits of inflammability, respectively.

Ingot. A solidified steel form made by pouring molten steel from a steel furnace into a mould.

Inherent Ash. Incombustible mineral matter inseparable from the coal substance prior to combustion. In the course of the mechanical cleaning of coals none of the inherent ash can be removed. See **Adventitious Ash; Ash.**

Inherent Moisture. Moisture in the pores of coal which cannot be removed by mechanical means; inherent moisture increases with decrease in rank, and coals of the lowest rank may contain from 15 to 16 per cent. See **Free Moisture.**

Inhibitors. Chemical substances added to oil to check or retard the occurrence of undesirable properties.

Injector. A device for feeding water into a boiler which is under pressure; the system utilizes a steam jet and a suitable arrangement of nozzles. Its operation is based upon the conversion of some of the kinetic energy of the steam into pressure energy; about one pound of steam is required to inject ten pounds of water into the boiler. The injector is suitable for small boilers only and it cannot feed water with an initial temperature above 90° F (32° C). See **Boiler Feed Pumps and Injectors.**

Inside Mix Burner. *Oil burner*, q.v., in which the steam or air and oil impinge and mix within the burner and issue as a foam or fog. See **Outside Mix Burner.**

Installed Capacity. In respect of electricity generation, the total generating capacity for which a plant is designed, measured in kilowatts (thousand watts) or megawatts (million watts). Installed capacity is generally about 7 to 8 per cent greater than output capacity (measured in kilowatts or megawatts "sent out") because of the power consumed at the power station for lighting and auxiliary plant.

Installed Load Tariff. A *tariff*, q.v., for the supply of electricity consisting in its simplest form of a unit charge plus a standing charge related to the number of kilowatts of installed electrical load on a premises.

Institute of Fuel. A professional Institute founded in 1927 on the merging of the Institution of Fuel Economy Engineeers and the Institution of Fuel Technology. On the petition of The Institute, a Royal Charter of Incorporation was granted by H.M. King George VI on 12th August, 1946. The main object of The Institute is the "advancement of scientific knowledge in the preparation, treatment and utilization of sources of heat and power of all types in all applications". The Royal Charter authorizes a membership structure comprising Honorary Members, Fellows, Members and Associate Members (all being known as corporate members and "Chartered Fuel Technologists") and several non-corporate grades of membership. Membership numbers over 6000.

Insulating Firebrick. A fireclay product of high porosity, which withstands high temperatures and yet possesses good heat insulating value. Of low heat storage capacity it enables a furnace to be brought quickly up to its working temperature; however because of the porosity of the brick it can only be used under "clean heat" conditions.

Insulation. The use of materials of low thermal conductivity in order to retain heat within a furnace or flue. See **Conduction.**

Integrator. An instrument which adds up, or integrates, the momentary rate of flow to give a total quantity of whatever is being measured over any period. Instruments are available, for example, to count or integrate the total quantity of steam produced or water evaporated.

Inter-bank Superheater, See **Superheater.**

Intermittent Kiln. In the heavy clay industry, a non-continuous kiln, the two main types being: (a) the rectangular; (b) the round or "bee-hive" kiln. Each kiln is normally connected to one or two external stacks; some round kilns have a stack in the centre which projects through the crown. They are fired by coal or oil. See **Brick Kiln.**

Intermittent Vertical Retort. A closed chamber used for the manufacture of *town gas*, by the *carbonization*, qq.v., of coal. A retort of this type is usually about 22 ft high and 10 ft wide; the depth varies from 8 in. at the top to 12 in. at the bottom. It holds about $3\frac{1}{2}$ tons of coal. The carbonization period lasts between 10 to 12 hours, including two hours steaming at the end. Heat is provided by producer gas.

Internal Water Softening. The softening of water inside a boiler. Sodium carbonate added to boiler water at pressures up to about 200 lb/in² will dissolve to give carbonate ions which will combine with calcium and magnesium ions as a loose, soft sludge. For higher pressures, sodium phosphate may be used; it is stable and does not add carbon dioxide to the steam. The phosphate can be supplied in various forms such as trisodium phosphate, disodium hydrogen phosphate and Calgon.

International Atomic Energy Agency. A United Nations agency whose main objective is "to accelerate and enlarge the contribution of atomic energy to peace, health and prosperity throughout the world." The statute for the International Atomic Energy Agency was approved on the 26th October, 1956, at an international conference attended by representatives of 82 countries held at the United Nations headquarters, New York. The statute came into force on the 29th July, 1957. The first session of the Agency was held in Vienna, Austria, in October, 1957. The Agency has at present about 76 members.

International Coal Classification (E.C.E.) A classification system for coals prepared by the Classification Working Party of the Coal Committee of the Economic Commission for Europe. It is a two-dimensional scheme, the horizontal parameter being based upon *volatile matter*, q.v., content up to 33 per cent; and on *calorific value*, q.v., for coals with more than 33 per cent of volatile matter on the *dry ash-free*, q.v., basis. This parameter divides the coals into 9 classes; the classes are then divided into groups and sub-groups according to the coking or caking properties of the coals. See **Coal Classification Systems.**

International System of Units (SI)

International System of Units (SI). A metric system adopted in 1954 by the Tenth General Conference on Weights and Measures from the basic group of units from which could be derived in a coherent and rational way practically all the units required by the physical sciences and engineering technology. The six defined units are:

Length	Metre
Mass	Kilogramme
Time	Second
Electric current	Ampere
Thermodynamic temperature	Degree Kelvin
Luminous intensity	Candela

At the Eleventh General Conference in 1960 it was formally resolved that this system be designated as "The International System of Units" with the symbol (SI). The system has also been accepted by the International Organisation for Standardisation (ISO). The system provides for supplementary and derived units, e.g. for velocity, the metre per second; for force, the newton ($N = kg\,m/s^2$). See **British System; Metric System; M.K.S. System.**

International Temperature Scale. A temperature scale adopted in 1948 by the General Conference on Weights and Measures; temperatures on this scale are designated as °C (Celsius). The international temperature scale is based on a number of fixed and reproducible equilibrium temperatures (fixed points) to which numerical values are assigned. The fundamental points are:

	°C
Melting point of ice	0
Boiling point of water	100

The primary points are:

Boiling point of oxygen	−182·970
Boiling point of sulphur	444·600
Freezing point of silver	960·8
Freezing point of gold	1063·0

There are also a number of secondary fixed points. See **Temperature Scales.**

Interruptible Contract. A contract for the supply of *natural gas*, q.v., to boiler plant which permits an interruption of supply, if necessary, to meet the domestic home heating demand which has priority; a common feature of United States fuel contracts in

134

industrial areas. Firms likely to be affected require alternative fuels such as oil or coal as "second fuels" during the winter months.

Inter-tube Superheater. See **Superheater.**

Inverse Square Law. In relation to radiant heat, a law governing the variation of heat received per unit area (λ) from a given source at varying distance (d); the heat received at any point varies inversely with the distance squared:

$$\lambda = \frac{1}{d^2}$$

This means that equal areas of material held at distances of 2, 7 and 10 ft from a source of radiation will receive energy at $\frac{1}{4}$th, $\frac{1}{49}$ and $\frac{1}{100}$ of the rate of the heat received at a distance of one foot from the source. The law also applies to light waves. In air pollution control, the ground level concentration of a pollutant varies inversely with the square of the *effective height of emission*, q.v.; hence a doubling of the effective height reduces ground level concentrations by 75 per cent.

Inversion. A temperature inversion in the atmosphere, in which the temperature, instead of falling, rises with height above the ground. With the colder and heavier air below, there is no tendency to form upward currents and turbulence is suppressed. Inversions are often formed in the late afternoon when the radiation emitted by the ground exceeds that received from the sinking sun. Inversions are also caused by katabatic winds, i.e. cold winds flowing down the hillside into a valley, and by anticyclones. See **Lapse Rate; Subsidence Inversion.**

Ion. An *atom*, q.v., or molecule which has lost or gained one or more electrons, thus possessing a net positive or negative charge. The loss of electrons produces a positive ion (cation); the gain of electrons produces a negative ion (anion). See **Electron.**

Ion Exchange Process. A method of softening water which depends on the property of certain synthetic resins which enables them to give up ions in exchange for ions from the water; thus calcium and magnesium cations can be replaced by sodium cations. When the resin has given up most of its sodium ions, the efficiency of the process begins to fall and the exchange material has to be regenerated. To regenerate the resin a strong solution of brine is used from which sodium ions are given up to the resin. Calcium and magnesium ions take their place and remain in solution with the chloride anions. See Fig. 13. Also known as the base exchange process. See **Lime Soda Process.**

Ionization

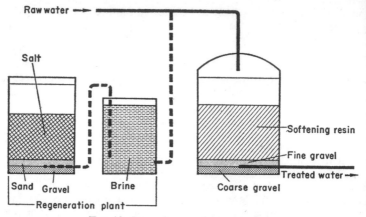

FIG. 13. *Ion or base exchange process.*

Ionization. The process by which a neutral atom or molecule loses or gains electrons, thus becoming electrically charged. See **Ion.**

Ionization Chamber. A device for measuring the quantity of ionizing radiation consisting of a chamber containing two electrodes and a gas. The radiation is measured by collecting the ions which it produces in the gas at the electrodes under the action of an apppplied electric field. See **Radiation Detector.**

Ion Pair. A positive *ion*, q.v., and a negative ion having charges of the same magnitude but opposite sign formed by the *ionization*, q.v., of a neutral atom or molecule.

IP Burning Test. A test for the burning quality of kerosine. The deposit-forming tendency of a kerosine is determined by the char formed on a wick during a 24 hour test.

Irradiation. The exposure of matter to radiation.

Isenthalpic. Constant *enthalpy*, q.v.

Isentropic. Constant *entropy*, q.v.

Isobaric. Constant pressure; also called isopiestic.

Isobutane, C_4H_{10}. A specific hydrocarbon compound which reacts with light olefins in the presence of sulphuric or hydrofluoric acid to make high octane alkylate gasoline.

Isochoric. Constant volume; also called isometric.

Isokinetic. In respect of taking a sample of flue gas or other exhaust product, a situation in which the flow of gas into the *sampling probe*, q.v., has the same flow rate and direction as the gas being sampled; hence, "isokinetic sampling".

Isomerization. An oil refinery process for producing isobutane from *n*-butane, or isopentane from pentane. In respect of butane an aluminium chloride catalyst is employed. Liquid *n*-butane is passed through a drying tower, vaporized and passed through a reactor containing the catalyst where approximately 40 per cent of the *n*-butane becomes isobutane.

Isopentane. C_5H_{12}. A component of high grade gasoline, it is obtained in practically its pure form by highly efficient fractionation of light straight run gasolines.

Isothermal. Constant temperature.

Isotopes. Atoms possessing nuclei with the same number of protons in each but with different numbers of neutrons. Isotopes are distinguished from one another by means of their mass number, which is the sum of the numbers of protons and neutrons in the nucleus.

J

Jerk System. A category of *fuel injection equipment*, q.v., for diesel engines; separate fuel pumps are used to deliver fuel to each cylinder during the injection period only. See **Diesel Engine.**

Jet Condenser. A steam condenser in which cooling is achieved by mixing the steam with a spray of water.

Jones Oil Gasification Process. A cyclic process for the manufacture of gas in which oil is thermally cracked with steam to produce a gas having a calorific value in the range 360 to 520 Btu/ft³. The process yields appreciable quantities of carbon black about half of which is used in the process itself. The plant consists of a two vessel generator and two filter gas producers. One blow and run cycle normally take 7 to 10 minutes and then the flow is reversed. During the blow, primary air is blown through one of the gas producers, the resultant producer gas being burned with preheated secondary air above and through the chequer work in the generator; the waste gas is discharged to atmosphere. Steam is then introduced through the gas producer, the resulting blue water gas passing to the generator; oil is sprayed into the generator and the oil vapours crack in the atmosphere of blue water gas and steam; half the suspended carbon black is filtered out in the other filter gas producer which contains a bed of mechanically agitated refractory spheres; this carbon black is consumed when the blow and run cycle is repeated from the opposite direction. The remaining carbon is removed in a wash box.

Joule. Unit of energy, including work and quantity of heat; the work done when the point of application of a force of 1 newton is displaced through a distance of 1 metre in the direction of the force.

$$
\begin{aligned}
1 \text{ joule} &= 1 \text{ watt second} \\
&= 10^7 \text{ erg} \\
&= 0 \cdot 238846 \text{ calorie} \\
&= 0 \cdot 526565 \times 10^{-3} \text{ Chu} \\
&= 0.947817 \times 10^{-3} \text{ Btu} \\
&= 0 \cdot 737562 \text{ foot pound force} \\
&= 1 \text{ newton metre}
\end{aligned}
$$

See **Dyne**; **Erg**.

Joule's Equivalent. The amount of mechanical energy which can be transformed into one heat unit; it is equivalent to 778·2 ft lb/Btu.

Junkers Calorimeter. A *gas calorimeter*, q.v., in use in Europe and the United States; the *calorific value*, q.v., of a known volume of gas is calculated from the increase in temperature of a measured volume of water heated by the gas under test.

K

Kaldo Furnace. A *steelmaking furnace*, q.v., of Swedish origin; it consists of a cylindrical furnace lined with refractory bricks and as it rotates a jet of oxygen is blown against the surface of the molten metal in the vessel.

Katabatic Winds. See **Inversion**.

Kelvin Scale. A temperature scale based on three fixed points: (a) the lowest (0°) being absolute zero; (b) the second (273·15°) being the temperature of melting ice; (c) the highest (373·15°) being the boiling point of water at normal atmospheric pressure. The Kelvin and Rankine scales are absolute scales, $0^\circ \text{ K} = 0^\circ \text{ R}$; 1 Kelvin degree = 1·8 Rankine degrees. See **Temperature Scales**.

Kerosine. A petroleum fraction with a boiling range of 302 to 572° F (150 to 300° C) and specific gravity of 0·78 to 0·82. It is used in lamps and stoves, and as a fuel for jet aircraft.

KeV. Symbol for 1 kilo-electron volt = 10^3eV.

Kilogramme Calorie (kcal). The amount of heat required to raise one kilogramme of water through one degC; equal to 1000 gramme calories. See **Calorie**.

138

Kilovolt-ampere, kVA. A unit applying to *alternating current*, q.v., only; it is the product of the pressure in voltage and current in amperes which, when multiplied by the *power factor*, q.v., gives the power in kilowatts. It is sometimes referred to as the "apparent" power.

Kilowatt. One thousand watts; the larger unit of electrical power. In respect of a *direct current* supply, or a single phase *alternating current*, supply at a *power factor*, qq.v., of unity, the number of kilowatts is obtained by multiplying the pressure in volts by the current in amperes and dividing by 1000. If the power factor is less than unity, the answer obtained must be multiplied by the power factor. In three-phase systems the voltage across the phases, current per phase, and the power factor must be multiplied together and further multiplied by the square root of 3 (i.e. 1·73) to give an answer in kilowatts.

Kinematic Viscosity. The ratio of the *dynamic viscosity*, q.v., to the density of a liquid, the units being stokes or *centistokes*, q.v. Kinematic viscosities are quoted at 122° F (50° C). See **U-tube Viscometer; Viscosity.**

Kinetic Energy. The energy possessed by a body because it is moving.

Kinetic Plume Rise. The rise of a chimney plume caused by the upward velocity of the gases alone. A formula suggested by Nonhebel (*Gas Purification Processes*, Newnes, 1964) for the calculation of this rise is:

$$h_m = \frac{4 \cdot 64 \, (VQ)^{1/2}}{u(1 + 0 \cdot 67u/V)} \text{ ft}$$

where, Q = gas rate measured at temperature T, ft^3/s

V = efflux velocity, ft/s
u = wind speed, ft/s
T = absolute temperature at which the gas has the same density as the ambient air (°K.)

King-Maries-Crossley Formula. Or "K.M.C. formula", a correction formula for determining the *mineral matter*, q.v., content of coal from laboratory data (*J. Soc. Chem. Ind.*, 1936, **55**, 277T). This formula was revised in 1956 by the *National Coal Board*, q.v., and in its revised form states:

$$M.M. = 1 \cdot 13A + 0 \cdot 5 \, S_{pyr} + 0 \cdot 8 \, CO_2 - 2 \cdot 8 \, S_{ash} + 2 \cdot 8 \, S_{sulph} + 0 \cdot 5 \, Cl$$

where, as percentages of the air-dried coal,

$M.M.$ = mineral matter in the coal
A = ash obtained on incineration
S_{pyr} = pyritic sulphur in the coal
CO_2 = carbonate CO_2 in the coal
S_{ash} = total sulphur in the ash
S_{sulph} = sulphur present as sulphates in the coal
Cl = total chlorine in the coal.

A simplified formula subsequently derived by the British Coal Utilisation Research Association is:

$$M.M. = 1 \cdot 10A + 0 \cdot 53\,S + 0 \cdot 74\,CO_2 - 0 \cdot 32$$

where, $M.M.$ = mineral matter
A = ash
S = total sulphur
CO_2 = total carbon dioxide.

Kirchoff's Radiation Law. A statement that the radiating capacity of a body is proportional to the absorbing capacity of the body, at any given temperature and any given wavelength. A perfect emitter and absorber is a *black body*, q.v., with an emissivity and absorbtivity of unity.

Knock Limited Power. The maximum power an engine will produce under a prescribed set of conditions without "knocking", q.v.

Knocking. Detonation in engine cylinders, accompanied by a knocking sound, associated with decrease in power output and overheating; combustion in a cylinder should proceed in a smooth manner at a moderately fast rate, not instantaneously. Also called "pinking".

Koppers-Hasche Process. A cyclic process for reforming methane and natural gases to make them more suitable for use as town gas; the process consists of partial oxidation of the gases with air using an alumina catalyst:

$$2CH_4 + O_2 \rightarrow 2CO + 4H_2$$

The process is carried out in two catalyst packed horizontal chambers, reversal of flow taking place every minute. The amount of air used determines the *calorific value*, q.v., of the gas produced.

Krypton. Kr. An inert gas; at. no. 36, at. wt. $83 \cdot 80$; it is an important product of *fission*, in a *nuclear reactor*, q.v.

L

La Mont Boiler. A *water-tube boiler*, q.v., employing forced circulation. The boiler heating surface comprises a number of tube elements working in parallel, the inlet end of these tube elements being expanded into the distribution header of the boiler. Controlled circulation by pump affords protection against the overheating of boiler pressure parts; it permits the use of small bore tubes which increases the efficiency per square foot of tube surface. The La Mont design has been adopted for all classes of plant ranging from low pressure water boilers up to large power station units incorporating all types of firing equipment, and utilizing all classes of fuel. See Fig. 14.

Lagging. Insulating material applied to the external surfaces of boilers, flues, pipes and equipment to reduce heat losses.

Lambert's Law of Radiation. A statement that the radiation from a hot surface in a direction at an angle with the surface varies as the cosine of the angle ϕ between the direction of the radiation and the normal to the surface. The radiation in a given direction is therefore:

$$q = q_n \cos \phi$$

where q_n is the radiation normal to the surface. See Fig. 15.

Lancashire Boiler. A horizontal shell boiler, the shell containing two plain or corrugated furnace tubes running through the boiler from end to end; it has a brick flue setting. The hot combustion gases leave the combustion zone and pass through the furnace tubes to the rear of the boiler where they enter a downtake in which a superheater is sometimes located. From the downtake the gases enter a bottom flue and return again to the furnace front; there they divide into two streams passing down side flues, one being situated on each side of the boiler, to the rear. Here the gases reunite in a single main flue and pass to the chimney. The working pressure of a Lancashire boiler rarely exceeds 250 lb/in². Fitted with a feedwater economizer and mechanically fired the thermal efficiency may be raised to about 75 per cent. Evaporative capacities range from about 5000 to 15,000 lb/h "f. and a. 212° F". The Lancashire boiler is being superseded by economic boilers and package boilers of higher efficiency. See **Boiler; Cornish Boiler.**

Lapse Rate. The rate at which temperature decreases with altitude; the vertical temperature gradient of the atmosphere. *Adiabatic,*

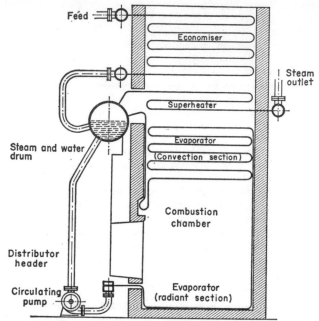

FIG. 14. *La Mont forced circulation water-tube boiler.*

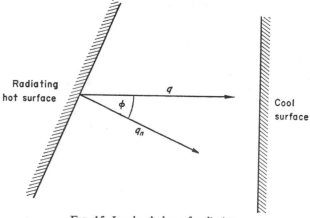

FIG. 15. *Lambert's law of radiation.*

q.v., temperature changes proceed at a definite rate. The dry adiabatic lapse rate is $5.4°$ F ($3°$ C) per thousand feet. The saturated adiabatic lapse rate is about half the dry adiabatic lapse rate. Average lapse rates lie between the two. At times, the temperature of the atmosphere, instead of falling, increases with height; this is known as an inversion. These temperature gradients and their characteristic climatic conditions are important factors in determining the rate of diffusion of air pollutants in the atmosphere. See **Air Pollution; Inversion; Static Stability; Superadiabatic Lapse Rate.**

Large Coal. Coal with no specified maximum size, but with a specified minimum size of say 3 in. or more.

Latent Heat. Hidden heat; the number of heat units absorbed or released by unit weight of a substance during a change of physical state, e.g. from water into steam, or from water into ice. The heat transfer involved is not shown by a thermometer, hence the description "hidden" heat. See **Latent Heat of Fusion; Latent Heat of Vaporization.**

Latent Heat of Fusion. Heat required to melt unit mass of a substance while its temperature remains constant; 144 Btu (80 Chu) are required to melt one pound of ice.

Latent Heat of Vaporization. Heat required to convert unit mass of a liquid into the gaseous state while its temperature remains constant. The latent heat of steam is 972 Btu (540 Chu) per pound of water, at atmospheric pressure.

Lattice. In a *heterogeneous reactor*, q.v., a geometrical arrangement of the fuel elements and the *moderator*, q.v.

Lead. Pb. A heavy metallic *element*, q.v., at. no. 82, at. wt. 207·19. It is widely used as a radiation shielding material because of its high density. See **Gasoline Additives.**

Lead Peroxide Candle. An instrument for measuring the relative concentrations of *sulphur dioxide*, q.v., in the general atmosphere. It consists of exposing a lead peroxide surface; the rate of sulphation is proportional to the sulphur dioxide. The instrument is usually exposed for a period of one month. The casual errors are relatively large and the instrument may be affected by humidity; the relationship of the readings expressed in mg/SO_3 100 cm^2 day, to actual concentrations of SO_2 is only approximate. See **National Survey of Air Pollution; Volumetric SO_2 Apparatus.**

Lead Susceptibility. The degree of response by way of an increase in *octane number*, q.v., shown by a gasoline to which tetraethyl lead (TEL) has been added; the response is not uniform but depends on the composition of the gasoline, e.g. the presence of sulphur compounds reduces lead susceptibility.

Lean Gas. Gas of relatively low calorific value. See **Rich Gas.**

Lean Oil. *Absorption oil*, q.v., from which dissolved natural gasoline fractions have been removed.

Leaving Loss. The loss of energy in the velocity of turbine exhaust steam.

Lever or Steelyard Safety Valve. A type of *safety valve*, q.v., which possesses one weight only equal to the maximum load when the weight is at the end of the lever. The weight should be attached in such a way that it cannot be moved inadvertently.

Light Ends. The lower boiling components of a mixture of hydrocarbons.

Light Meter. A direct reading instrument for measuring illumination, incorporating a photoelectric cell and an indicating meter calibrated in lumens per square foot, or in foot-candles.

Light Sensitive Pyrometer. A device for measuring temperatures, utilizing a light sensitive element such as a *photoelectric cell*, q.v., to measure the intensity of light emitted by a hot object. The magnitude of current through the cell is determined by the intensity of radiation; the current is amplified electrically and recorded. The instrument is suitable for temperatures of 1290° F (700° C) upwards. See **Radiation Pyrometer.**

Lignite. A solid fuel intermediate between *peat*, q.v., and other coals; it shows a definitely woody structure. The carbon content is lower than for *brown coal*, q.v., while the moisture content may be as high as 50 per cent. Lignite is extensively used in many countries for heating and steam raising.

Lime-base Grease. A grease with water-resistant properties used for lubrication under wet or moist operating conditions.

Lime Kiln. In its most usual and modern form a vertical steel cylinder lined with firebrick. Heat is supplied from the combustion of coal, oil or producer gas. Limestone is crushed and loaded continuously into the top of the kiln, the lime being withdrawn from a hopper at the base; each lump of limestone is in the kiln several hours gradually falling towards the hot zone which has a temperature of about 2000° F (1100° C). During the process, carbon dioxide is driven off from the stone, the result being quicklime; the chemical reaction which takes place is:

$$CaCO_3 \rightarrow CaO + CO_2$$

The problem of completing the combustion process in an atmosphere consisting largely of nitrogen and carbon dioxide gives rise to *dark smoke*, q.v., emissions. Limestone is also calcined or burned in rotary kilns, similar to those used in the cement industry.

Lime Silica Block. An insulating material consisting of reacted hydrous calcium silicate. It is used on plant and pipes usually below 1200° F (649° C).

Lime Soda Process. A method of softening water in which most of the calcium and magnesium are removed by controlled additions of lime and soda ash; the unwanted calcium and magnesium salts fall out as a sludge, the softened water being filtered from the top. Lime soda treatment needs a hard water of more than 100 p.p.m., but with not too high a proportion of magnesium. The process will not remove all the hardness, however, and must be followed by the *ion exchange process*, q.v.

Line Reversal Method. A method of measuring the temperature of clear flames. To obtain visible spectrum lines it is necessary to add a metal compound such as a sodium salt; this is done in the sodium line reversal method. The emission spectrum of a flame containing salt shows the two yellow D lines of sodium in the emission. If a bright source emitting a continuous spectrum is viewed through the flame, the D lines appear as dark lines against the continuum. If the brightness of the background is adjusted then there is a condition at which the sodium lines just disappear. Under prescribed optical conditions the brightness temperature of the background and the flame temperature are equal. The brightness temperature of the background source is determined by a separate measurement using a disappearing-filament optical pyrometer. It has been possible by this method to make measurements as high as about 9000° R (5000° K). See **Pyrometer; Thermometer.**

Linz-Donawitz Converter. A *steelmaking furnace*, q.v., of Austrian origin utilizing oxygen; an L-D vessel resembles a normal *Bessemer converter*, q.v., but without provision for air injection in the base. After charging with scrap steel and molten iron, a long copper-tipped lance is lowered automatically until its nozzle is 2 to 3 ft above the surface of the metal. For about 18 minutes a jet of oxygen with a velocity of 2000 ft/s is projected down into the molten metal. Steel is produced without the nitrogen brittleness of normal basic Bessemer steel.

Liquefied Petroleum Gases (L.P.G). Paraffin hydrocarbon gases comprising propane, butanes and pentanes, obtained from natural gas wells and in petroleum refining. For efficient transportation, storage and use, propane and butane are liquified under pressure and distributed in cylinders—hence the description "bottled gas". Apart from domestic use these gases are used for the manufacture of chemicals and with other hydrocarbons used as components of aviation fuels. See **Butane; Pentanes; Propane**.

Liquid-in-Glass Thermometer. A device for measuring temperature; the temperature is indicated by the differential expansion of a liquid with respect to its glass container. The liquid may be mercury, alcohol, toluene or xylene, choice depending upon the temperature range over which the instrument will be used. For ordinary purposes mercury is suitable; if the temperatures to be measured are likely to fall below $-37.9°$ F $(-38.8°C)$ then ethyl alcohol or some other organic liquid is employed. The overall range for instruments of this type is $-328°$ F to $+932°$ F $(-200°$ C to $+500°$ C). See **Pyrometer; Thermometer**.

Liquid-in-Steel Thermometer. A device for measuring temperature; the measuring element consists of a Bourdon tube gauge acting as a pressure measuring device, the expansion of the liquid altering the shape of a hollow, flexible metal coil which is connected to an indicator. Mercury is the usual liquid. The instrument is suitable for the range 572 to 1292° F (300 to 700° C), although by using an inert gas lower temperatures can be reached satisfactorily. See **Pyrometer; Thermometer**.

Lithotype. A macroscopically recognizable banded coal type, greater than 3 mm thickness; it includes *clarain, durain, fusain,* and *vitrain,* qq.v. See **Microlithotype**.

Live Steam Accumulator. A vessel containing water in which surplus steam from the main steam generator is stored under a gradually rising pressure, and regenerated as steam under a falling pressure. A steam accumulator is one method of meeting fluctuating demands for steam; under ideal conditions it means that the main steam generator can be fired at a rate corresponding to the average steam consumption throughout the day. It is claimed that peaks may be met as high as 60 to 80 per cent above average demand.

Load Factor, Plant. Actual output expressed as a percentage of what would have been produced had all the plant been operating continuously throughout a specified period. In respect of electricity generating plant it is the average hourly quantity of electricity sent out during the year, expressed as a percentage of the average output capacity during the year. See **Availability**.

146

Load Factor, System. In relation to an electricity supply system, the ratio, expressed as a percentage, of the average load throughout the year to the highest load on the system. In order, for purposes of comparison, to allow for year to year weather variations, the *Central Electricity Generating Board*, q.v., adjusts the actual system load factor to a basis estimated on "average cold spell" conditions; on this basis the annual load factor is about 52 per cent. In the United States load factors of 60 to 65 per cent are achieved. In comparing system load factors in different countries it is necessary to take into account differences in consumer requirements determined by, among other things, social, climatic and economic conditions. System load factor is a measure of the fluctuating pattern of consumer demand, not of the efficiency with which supply industries use their plant.

Load/Rate Tariff. A *tariff*, q.v., for the supply of electricity in which the consumer pays a standing charge related to the demand he subscribes for, and when his demand is less than the subscribed demand the units consumed are charged at a low price; when his demand is above the subscribed demand the units of electricity in excess are charged for at a relatively high price. This form of tariff is used extensively in Norway.

Load, Simultaneous Maximum. On the British electrical power system, the maximum load actually met on the national grid for any half-hour together with the load at that time on any stations not tied into the grid. This may differ from the simultaneous maximum potential demand by any load shed through voltage reduction, disconnection or any reduction in frequency.

Load, System Peak. On the British electrical power system, the annual maximum simultaneous load; it usually occurs in the late afternoon of some cold working day in December or January.

Lock Hopper. A vessel attached to a pressure gasifier to permit the intermittent passage of solid fuel without direct communication between the gasifier and the atmosphere.

Locomotive Boiler. A type of *boiler*, q.v., used on steam locomotives; the boiler is fitted with an integral water-cooled firebox, the hot combustion gases passing from the firebox through a horizontal bank of fire-tubes situated within the shell of the boiler into a smoke box and out of a short chimney at the opposite end. The locomotive boiler is a relatively efficient steam generator. Combustion rates for coal may at times reach 100 lb/ft^2 h on the grate as a result of the induced draught obtained from the operation of the blast pipe. The latter is situated in the smoke box below the chimney; through it passes the exhaust steam from the cylinders, the velocity of the steam inducing draught through the firebox. While stationary, the creation of draught depends on the use of a "blower", consisting of a ring of steam jets at the base of the chimney which can be operated by a valve in the cab. When moving the blast pipe takes over. Despite the high thermal efficiency of the boiler, the overall efficiency of the locomotive is quite low; only a small percentage of the heat passed forward to the cylinders is converted into useful work. The general average efficiency of locomotives of modern design is about 7 per cent. A stationary locomotive boiler is a simple adaptation of the normal locomotive boiler; it is sometimes used where head room in a boiler house is limited and appears to have been widely used to power oil-drilling machinery in various parts of the world.

Log-Sheet. A detailed record of instrument readings and plant operating details kept by a plant operator or supervisor.

Long Flame Coal. *Coal* of high *volatile matter*, qq.v. See **Short Flame Coal.**

Looping. The behaviour of a chimney plume when the environmental eddies so influence the plume that the *efflux velocity*, q.v., or momentum and thermal buoyancy of the gases are ineffective and the plume zig-zags up and down. See **Coning.**

Los Angeles Smog. Smog of a photochemical nature, largely attributable to the effect of sunlight on *motor vehicle exhaust gases*, q.v. Basic to the Los Angeles control system is a programme of monitoring both weather conditions and levels of air contaminants. Three stages of alert related to the severity of exposure have been adopted, and appropriate control measures are instituted whenever necessary. The limits of contamination for each of the three stages are summarized in Table 7. In the first stage alert, all unnecessary activity which might pollute the air must be avoided. A second stage alert indicates a health menace, and the County Air Pollution Control Director may impose limitations on the general operation of vehicles and may restrict the operations of public utilities and other industries to those essential to continued operation of the industrial complex. The third stage alert is evidence of a dangerous health menace and appropriate measures may be taken under the California Disaster Act to limit activities to emergency needs.

Low-Heat Value (LHV). Synonymous with *net calorific value*, q.v.; a term most often used in the United States of America.

Low Ram Coking Stoker. A *coking stoker*, q.v., in which the ram or coal feeder is set low in the flue relative to the grate thus enabling

TABLE 7—AIR POLLUTION ALERT LEVELS IN LOS ANGELES COUNTY

	Stages of alert (Contaminant concentration in p.p.m.*)		
Contaminant	1st	2nd	3rd
Ozone	0·5	1·0	1·5
Nitrogen oxide	3	5	10
Carbon monoxide	100	200	300
Sulphur oxides†	3	5	10

*Parts per million by volume.

†Ambient air quality standards adopted by the California Department of Public Health, 4th December, 1959 classify an SO_2 level of 1 p.p.m. for 1 hour or 0·3 p.p.m. for 8 hours as "adverse" levels, i.e. levels at which there will be sensory irritation, damage to vegetation, reduction in visibility, or similar effects.

a wide ram to be fitted; this ensures a more even distribution of fuel across the grate than was possible with earlier high ram designs. A wide range of British coals can be burned on the low ram stoker; non-coking to medium coking coals are suitable. Burning rates do not generally exceed 40 to 45 lb/ft² h for this type of stoker.

Low Temperature Deposits. Deposits on the cooler parts of a boiler system such as economizers and air-heaters consisting of alkali chlorides, sulphur compounds and phosphates. Dry-bottom pulverized fuel-fired boilers are practically immune from "back-end" corrosion, except when the chlorine and sulphur in the coal are very high (chlorine exceeding 0·6 per cent). See **High Temperature Deposits.**

Lubricating Oil. A high boiling paraffin hydrocarbon; the properties usually required are: (a) stability at high temperatures; (b) fluidity at low temperatures; (c) moderate change only in viscosity over a broad temperature range; (d) sufficient adhesiveness to keep it in place under high shear forces. See **Lubrication.**

Lubrication. See **Boundary Lubrication; Hydrodynamic Lubrication; Lubricating Oil; Solid Lubricants.**

Lumen. Unit of quantity or total flux of light; the total amount of light necessary to illuminate one square foot of surface to the value of one *foot-candle*. In the *International System of Units*, qq.v., it is the unit of luminous flux; the flux emitted within a unit solid angle of one *steradian*, q.v., by a point source having a uniform intensity of one *candela*, q.v.

Lurgi Process. A process originally developed in Germany to gasify brown coal; it has subsequently been applied in Britain for the total gasification of poor quality coal. The process is carried out under conditions of high temperature and pressure in the presence of steam and oxygen. The gas produced is largely hydrogen and carbon monoxide; it is then purified by passing it with steam over a catalyst to convert the carbon monoxide and steam into carbon dioxide and hydrogen by the *water gas shift reaction*, q.v.; the carbon dioxide is then removed in a Benfield plant. The final gas has a calorific value of about 400 Btu/ft^3 (222 Chu/ft^3). One plant is at Westfield, Scotland; another at Coleshill, Birmingham, England.

Lux. Unit of illumination; an illumination of 1 *lumen*, q.v., per square metre.

M

McCashney Incinerator. A wood-waste incinerator consisting of a " bottle "-shaped brick or steel shell, lined with a standard siliceous firebrick having an alumina content of about 22 per cent. The waste material is burnt as it is blown into the incinerator tangentially, the combustion of most of the material occurring while it is in suspension. The larger particles and any solid waste burn on the base of the unit. Air is also admitted at the bottom of the incinerator. The volume of both over and under-fire air is regulated to ensure optimum combustion; temperatures of 1832 to 2552° F (1000 to 1400° C) are developed. The McCashney incinerator, originally evolved by Mr. R. McCashney of Victoria, Australia, was developed by the Division of Forest Products, C.S.I.R.O., for general industrial use; it is widely used in Australia for the disposal of sawdust and wood shavings at sawmills and other wood-processing plants.

Macerals. Elementary homogeneous microscopic constituents of coal; they include alginite, collinite, cutinite, exinite, fusinite, inertinite, micrinite, resinite, sclerotinite, sporinite, telinite, vitrinite.

Magazine Boiler. A hot water boiler equipped with a built-in hopper to hold fuel for up to 24 hours continuous operation; from the hopper, fuel gravitates on to the fire and the rate of burning is thermostatically controlled. Automatic magazine boilers will operate at an efficiency of 80 to 85 per cent and can be supplied as a single boiler up to a rating of 10,000,000 Btu/h (5,600,000 Chu/h) suitable for a building of say 2,500,000 ft³ for central heating purposes. See **Boiler.**

Magnesia, 85 per cent. An insulating material, composed of 85 per cent of hydrated light basic magnesium carbonate and 15 per cent of long-fibred asbestos as a binding agent; it is suitable for temperatures up to about 570° F (299° C).

Magnesite Brick. A *refractory*, q.v., containing 84 to 92 per cent of magnesium oxide.

Magnetic Separator. A magnetic device installed in a conveyor system to attract and remove any undesirable tramp iron in the coal.

Magnetohydrodynamic Generation. Direct electricity generation utilizing hot flue gases and a magnetic field. In the simplest form of MHD generation a hot, electrically conducting fluid at about 4500° F (2500° C) is expanded through a nozzle and passed along a duct at high velocity. A magnetic field acts in a direction at right angles to the axis of the duct and the electric currents induced in the flowing gas are collected on electrodes. Seeding of the gas with a trace of readily ionizing material such as caesium or potassium vapour enhances the electrical conductivity by many orders of magnitude; cost necessitates a high rate of seed recovery. Research by the Central Electricity Generating Board is continuing to develop the principle of MHD generation; if such a technique could be combined with conventional plant it would serve as a "thermodynamic topper" helping to improve power station efficiency. See Fig. 16.

Magnox. A magnesium alloy containing small quantities of beryllium, calcium and aluminium to reduce the possibility of rapid oxidation and fire; used in British nuclear reactors as a canning material for uranium fuel elements. Magnox melts at about 1184° F (640° C). See **Magnox Reactor.**

Magnox Reactor. A *nuclear reactor*, q.v., first designed and built at Calder Hall in the United Kingdom which became the basis of a "generation" of commercial nuclear power stations. The design uses uranium fuel with magnesium alloy (Magnox) as a canning material; graphite as a moderator and carbon dioxide as a coolant gas. See **Magnox.**

Main Stop Valve. A valve which enables a boiler to be isolated from the steam pipe main and permits regulation of the flow of steam from a boiler. It is fitted to the highest part of the boiler, as far as practicable from the water surface. An anti-priming pipe is usually fitted inside the boiler to the steam outlet to reduce the risks

151

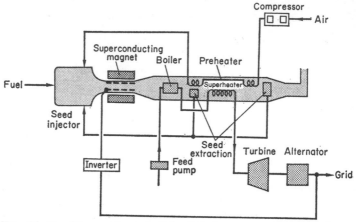

FIG. 16. *Possible arrangement of fossil fuel power station of the future, equipped with an MHD "topper" in addition to the normal steam turbo-generators.*

of *priming*, or *carry-over*, qq.v. For pressures up to 150 lb/in² the body of the stop valve is cast iron; above this pressure steel is used. The main stop valve incorporates a non-return valve if the boiler supplies steam to a common main.

Make-up Water. Water added to a boiler to compensate for the water lost as steam.

Malleable Cast-iron. *Cast-iron*, q.v., made from constituents which solidify "white", i.e. with all the carbon combined and none present as graphite; the "white" castings are annealed to break down the hard brittle cementite and produce malleable castings.

Manchester Process. A wet scrubbing process for the removal of *hydrogen sulphide*, q.v., from refinery and petroleum oil gas streams. The scrubbing medium is a suspension of ferric hydroxide, $Fe(OH)_2$, in a dilute solution of ammonia or sodium carbonate. Sulphur is recovered by blowing air through the solution. See **Hydrogen Sulphide Removal**.

Manifold. A piping arrangement which allows one stream of liquid or gas to be divided into two or more streams, or several streams to be collected into one stream, as in the case of the inlet and exhaust manifolds of an automobile engine.

Manometer. A *draught gauge*, q.v., utilizing a U-tube containing water.

Marine Boiler. A cylindrical shell boiler with flat ends, equipped with several furnace tubes; the gases flow through these furnace tubes to the rear of the boiler, returning to a "smoke box" at the boiler front through banks of small diameter fire-tubes. The boiler may be single- or double-ended, the latter presenting a "back-to-back" arrangement. Known as the Scotch marine boiler. See **Boiler**.

Mass Number. The total number of protons and neutrons in a *nucleus*, q.v., it is equal to the integer nearest in value to the atomic mass when the latter is expressed in atomic mass units. In the symbol for the nuclide, the mass number is indicated by a super-script following the element symbol; for example, U^{238}.

Mass Transfer. In respect of a fluid consisting of two or more components, in which a concentration gradient exists, the movement or flow of the components so as to reduce the concentration gradient. In a still fluid, or a fluid flowing streamline in a direction normal to the concentration gradient, mass transfer is effected by molecular diffusion, a slow process; in conditions of turbulence, molecular diffusion is supplemented by eddy diffusion, a much more rapid mixing process.

Mature Coal. A general term for high rank *coal*, q.v. See **Immature Coal**.

Maximum Demand Tariff. A *tariff*, q.v., for the supply of electricity which, in its simplest form consists of two parts, a unit charge and a charge levied on the number of kilowatts of electricity of maximum demand registered over, say, half an hour.

Mayer Curve. A graphical curve which gives directly, for any point and for any specific gravity, the percentage of ash in the "floats" of a *float and sink test*, q.v.; the cumulative ash content of the floats, expressed as a percentage by weight of the coal feed, is plotted against a cumulative percentage yield of floats over a range of specific gravity from 1·3 to 1·8 by increments of 0·1.

Mechanical Energy. Capacity for doing work. Mechanical energy takes either of two principal forms: (a) potential energy—the energy of position, (b) kinetic energy—the energy of motion.

Mechanical Power. The rate of expenditure of *mechanical energy*, q.v.; it is measured in foot-pounds per second or per minute. Machines are rated in horsepower where one hp is a rate of 550 foot-pounds per second. One *British Thermal Unit*, q.v., is the equivalent of 778 foot-pounds of mechanical energy.

Mechanical Rectifier. A transformer-rectifier for converting the alternating current of normal electricity supply into high tension direct current electricity; this unit has found wide use in electrostatic precipitators. It takes the form of an insulated disc with contacts on its periphery, or an insulated arm with two contact blades with sliprings, rotated by a *synchronous motor*, q.v., so arranged that the high voltage peaks of the a.c. wave form are selectively commutated. The high tension current is picked up by jumping across an air gap from the a.c. contacts arranged around the insulated disc or insulated arms. Mechanical rectifiers are less costly than static types, essentially rugged, and the performance is largely unaffected by fluctuations in process conditions. They are however noisy, require radio and television suppressors and emit nitrous oxide fumes. See **Electrostatic Precipitator; Rectifier; Static Rectifier.**

Mechanical Stoker. A device which feeds coal or other solid fuel to a furnace automatically, displacing hand-firing methods. See **Chain Grate; Chain Grate Stoker; Coking Stoker; Low Ram Coking Stoker; Sprinkler Stoker; Travelling Grate; Trickle Feed Stoker; Underfeed Stoker.**

Medium Voltage. Means a voltage exceeding 250 volts, but not exceeding 650 volts under normal conditions.

Mega-electron-volt (MeV). One million electron-volts.

Megawatt-day. The unit used to express the burn-up achieved or energy extracted from a fuel element; 1 megawatt-day = 24,000 kilowatt-hours.

Meniscus. The curved surface presented by a liquid in a tube, e.g. a draught gauge. In taking readings it is usual to take the bottom of the meniscus with water, and the top of the meniscus with a liquid such as mercury.

Mercaptans. Organic compounds having the general formula R—SH, meaning that the thiol group, —SH, is attached to a radical such as CH_3 or C_2H_5. The simpler mercaptans have strong, repulsive, garlic-like odours; odours become less pronounced with increasing molecular weight and higher boiling points. Ethyl mercaptan, C_2H_5SH, is a liquid of nauseous odour although readily oxidized to ethyl disulphide by exposure to air. Mercaptans may be produced in oil refinery feed preparation units as a result of incipient cracking; the offensive gases are burnt in plant heaters. Mercaptans arising in cracking units are removed by scrubbing with caustic soda; the mercaptans being removed from the spent caustic soda by stream stripping and subsequently burned in a process furnace.

Merit Order. The operation of power stations connected to a grid according to their fuel costs. Thus aggregate demand at any one time is met by bringing into operation the plant with the lowest incremental fuel costs per unit sent out, allowing for location and transmission costs; the most efficient plant is in use as nearly continuously as possible. At the other end of the scale, daily peaks in the winter months are met by bringing into use, for only a few hours a year, the oldest time expired plant with very high fuel costs and, more recently, gas turbine plant.

Mesons. Unstable particles having masses intermediate between those of the *electron*, and the *proton*, qq.v.

Metallurgical Coke. A hard coke produced in a *coke oven*, q.v., for use in blast furnaces.

Methanation. A method of enriching a gas containing carbon monoxide and hydrogen by reacting these two constituents over a catalyst to produce *methane*, q.v. The methanation reaction is:

$$CO + 3H_2 \rightarrow CH_4 + H_2O$$

The reaction is exothermic and is carried out between 570 and 750° F (300 and 400° C). The catalyst is usually based on nickel.

Methane. CH_4. A non-toxic gas having a *calorific value*, q.v., of approximately 1000 Btu/ft^3 (556 Chu/ft^3). Methane is the principal constituent of *natural gas*, q.v., being found in the petroleum districts of the world and other important areas such as the North Sea. Methane may be liquefied by refrigeration to a temperature of $-258°$ F ($-161°$ C); in this state it is reduced to $\frac{1}{600}$ of its volume as a gas. Methane is also produced during the process of digestion in sludge digestion tanks at sewage works, and occurs in coal mines as "firedamp". See **Sewage Gas**.

Methane Tankers. Ships specially equipped to carry liquefied natural gas from Algeria to Britain (Canvey Island). The storage tank insulation consists primarily of prefabricated balsa panels with plywood facing; insulation of the sides is augmented by a layer of glass fibre attached to the cold face. The temperature of the liquefied gas is $-258°$ F ($-161°$ C); in this state it is reduced to $\frac{1}{600}$ of its volume as a gas and requires no special pressure to keep it liquid. For maximum operating economy, the cargo "boil-off" should not exceed 0·3 per cent per day of tank content; this is utilized to supplement the main boiler fuel. Two ships "Methane Princess" and "Methane Progress" deliver annually to Britain

700×10^3 tons of LNG or some 354×10^6 therms, the equivalent of 10 per cent of the gas load of England, Wales and Scotland. See **Natural Gas Pipelines.**

Metric System. A system of measurement based on the centimetre, the gramme and the second. It is often referred to as the "c.g.s. system". The smaller size of the units, compared with the *British System*, q.v., commends them to the scientist. Some of the relationships between the units of the two systems are:

$$2\cdot54 \text{ cm} = 1 \text{ in.}$$
$$453\cdot6 \text{ g} = 1 \text{ lb}$$
$$39\cdot37 \text{ in} = 1 \text{ metre}$$
$$2\cdot2 \text{ lb} = 1 \text{ kg}$$

See **Dyne; Erg; Fluidity; International System of Units; M.K.S. System.**

MeV. The symbol for one million electron volts ($1 \text{ MeV} = 10^6 \text{ eV}$).

Micrinite. A substance in coal derived from finely comminuted plant debris; a minor constituent of *clarain*, q.v., but a principal component of *durain*, q.v. It is a dull black material.

Microcurie. One millionth of a *curie*, q.v. (10^{-6} curie).

Microlithotype. Microscopic banded coal type with a minimum width of 50 microns; includes carbargilite, clarite, clarodurite, durite, duroclarite, fusite, sporite, vitrinerite and vitrite. See **Lithotype.**

Micromerograph. A fast accurate instrument for determining the particle size distribution of powdered materials in the sub-sieve size range. It uses the principle of sedimentation in still air at atmospheric pressure. A cloud of particles is introduced at the top of a sedimentation column. These particles all fall the same distance at their terminal velocities on to the pan of a recording balance; a cumulative weight curve plotted against particle diameter is obtained directly from the recorder chart. It is claimed that particle size distributions of dry powders with particle size ranges from 1 to 250 microns are obtained easily from the micromerograph employing this technique.

Micromho. The electrical unit of *conductance*, q.v.

Micron (μ). A unit of length. It equals 1×10^{-3} mm; 1×10^{-4} cm; $3\cdot9 \times 10^{-5}$ in.

Middlings. A mixed material of adhering coal and dirt.

Mill Differential Pressure. The difference in air pressure in inches water gauge between the inlet and the outlet of a pulverized fuel mill.

Mol

Millicurie (mc). 1 mc = 10^{-3} *curie*, q.v.

Milliroentgen (mr). 1 mr = 10^{-3} *roentgen*, q.v.

Mineral Matter. In coal, clay or shale mixed with varying proportions of free silica, silicates and other compounds of iron, calcium, magnesium, titanium, and alkali materials. See **King-Maries-Crossley Formula.**

Mineral Wool. An insulating material prepared from molten slag, glass or rock, blown into fibres by steam or air jet or spun by high-speed wheels. It may be supplied as a mineral wool base block, consisting of mineral wool fibres and clay, moulded under heat and pressure; or as mineral wool blanket, the fibres being compressed into blanket form held in shape between wire mesh or expanded metal lath. It is claimed that the temperature limit for block is 1800° F (982° C), and for blanket 1000° F (538° C).

Ministry of Power. Set up in June, 1942, when it absorbed the former Mines and Petroleum Departments and the functions of the Board of Trade in relation to gas, electricity and later iron and steel, the United Kingdom Ministry of Power deals with policy considerations affecting the coal, oil, gas, electricity and iron and steel industries and is responsible for the general administration of the statutes dealing with those industries. The Ministry is also responsible under the Nuclear Installations (Licensing and Insurance) Act, 1959, for the siting, design, construction, operation and maintenance of nuclear power stations and other nuclear installations with special reference to safety.

M.K.S. System. A system of measurement based on the metre, kilogramme and second. It is an attempt to create units of a useful size while retaining the advantages of the *metric system*, q.v. See **British System; International System of Units.**

Moderator. A material used in a *nuclear reactor*, q.v., to slow down the neutrons by means of scattering collisions with the nuclei of the moderator. Moderators include *hydrogen*, (in the form of light water), *deuterium* (in the form of heavy water), *beryllium* and *carbon*, q.v., in the allotropic form of graphite. See **Neutron.**

Moisture. See **Bed Moisture; Free Moisture; Inherent Moisture.**

Mol. The quantity of a substance whose weight in pounds or grammes is numerically equal to its molecular weight. If expressed in pounds, it is called the pound mol; if in grammes, the gramme mol. The molal volume of any gas at normal temperature and pressure (0° C and 760 mmHg) is 22·412 litres/g mol, or 359 ft³/lb mol. The molal volume of any ideal gas at a particular pressure and temperature is the same; the molal volume of liquids and solids varies depending on their chemical nature and density.

Molecular Formula. A chemical formula which indicates the actual numbers of the constituent atoms in the molecule, as distinct from an *empirical formula*, q.v., which indicates the proportions only in which the constituent atoms are present in the molecule. For example, normal butane with an empirical formula of C_2H_5 has a molecular formula of C_4H_{10}. See **Constitutional Formula; Graphical Formula.**

Mollier Chart. A diagram in which *total heat*, or *enthalpy*, q.v., is plotted against *entropy*, q.v.

Mond Gas. A gas obtained by passing air and a large excess of steam over small coal at about 1200° F (650° C).

Motor Method. A test to determine the *octane number* of a *gasoline*, qq.v., utilizing an engine speed of 900 rev/min and a mixture temperature of 300° F (149° C). The conditions correspond fairly closely to the operation of car engines at high speeds and under heavy loads. It is a more severe test than the *research method*, q.v.

Motor Spirit. *Gasoline*, q.v., or petrol of various degrees of *volatility*, q.v., distilling within the range of about 86 to 392° F (30 to 200° C).

Motor Vehicle Exhaust Controls. Techniques for reducing the emission to atmosphere of carbon monoxide and hydrocarbon gases from the tail exhausts of gasoline powered motor vehicles. The General Motors method consists of injecting a certain amount of air under pressure into the exhaust port from each cylinder. Air is drawn from the air filter by a positive displacement pump and discharged to individual nozzles located in the manifold near the exhaust valve of each cylinder; the exhaust system is insulated. Before the gases leave the tail pipe, the combustion of carbon monoxide and partly burned hydrocarbons is completed. The air pump is driven by a belt from the crankshaft. The method has been approved by the California Motor Vehicle Pollution Control Board. See **Blow-by: Motor Vehicle Exhaust Gases.**

Motor Vehicle Exhaust Gases. The complex gases emitted from the exhausts of motor vehicle engines. In the *gasoline engine*, q.v., some 15 pounds of air is required per pound of fuel to ensure complete combustion; however, maximum power is achieved at a lower air fuel ratio of 12·5 to 1 and complete combustion does not take place. The result is the emission of substantial quantities of carbon monoxide together with other products of incomplete combustion such as alcohols, aldehydes, ketones, phenols, acids, esters, epoxides, peroxides and other oxygenates. Factors other than the air/fuel ratio also influence the composition and volume of exhaust emissions. The *diesel engine*, q.v., operates with an excess of air and very little unburned fuel is normally exhausted; smoke emission is associated with engine overloading and poor engine maintenance. It emits much less carbon monoxide and hydrocarbons than the gasoline engine, but somewhat more *nitrogen oxides*, q.v., and aldehydes. See **Motor Vehicle Exhaust Controls.**

Mott's Classification. A system of coal classification using *calorific value*, and *volatile matter*, as the principal coordinates, on a *dry-mineral-matter-free*, qq.v., basis. See **Coal Classification Systems.**

Mouthpiece. A cast-iron unit bolted to the back of the front plate of a boiler furnace tube; it gives access to the furnace for fuelling and other operations, but is normally closed by a fire-door. The space at the back of the mouthpiece is fitted with a firebrick liner to prevent the front and the boiler seams from becoming overheated.

Multi-cellular Collector. A *dust arrester*, q.v., consisting of a large number of very small cyclones arranged in parallel; the individual cells may be vertical, horizontal or sloping. High collecting efficiencies are claimed over a wide range of particle sizes, although below those obtainable with the *electrostatic precipitator*, q.v. See **Cyclone.**

Multi-gas Burner. A gas burner invented by L. T. Minchin based on the principle that very stable flames are formed when gas burns at a row of orifices lying along the ridge of a wedge-shaped cavity in the underside of the burner head; the flames are fan-shaped. In addition, the burner is constructed in such a way that a slow-moving stream emerges at certain critical points at the base of the flame, giving even greater stability for the flame formed by the main stream of fast-moving gases. Thus modified, it is claimed that the burner will give good stable flames even using Schlochteren methane, which contains 10 per cent nitrogen, and will not light back even on gases containing a high percentage of hydrogen. To switch from one gas to another it is only necessary to change the injector nipple, so that the rate of heat input is maintained constant. See Fig. 17.

Multi-gas Burner

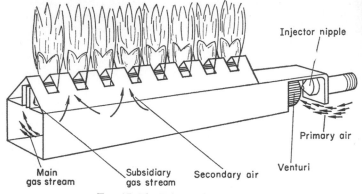

Injector nipple

Primary air

Venturi

Main gas stream

Subsidiary gas stream

Secondary air

FIG. 17. *The Tekni multi-gas burner.*

N

Naphtha. A light petroleum distillate with final boiling point up to 428° F (220° C).

Naphthene or Cyclo-paraffin Series. C_nH_{2n}. Cyclic *hydrocarbons*, q.v., in which the carbon atoms, instead of being linked in an open chain, form a closed ring; each carbon atom is joined to two carbon atoms and two hydrogen atoms and the compound is therefore "saturated". The ring may contain various numbers of carbon atoms; those most frequently found in petroleum contain five, six or seven. The rings may have side-chains attached to them consisting usually of groupings of carbon and hydrogen atoms. Naphthene or cyclo-paraffin compounds include cyclo-hexane, C_6H_{12}, and cyclopentane, C_5H_{10}. Being saturated, these compounds are chemically stable. The carbon/hydrogen ratio is between those of paraffins and aromatics.

National Air Sampling Network (NASN). An air pollution sampling network established in the United States in 1953. The NASN is operated in cooperation with health departments, air pollution control agencies, and other local organizations. The basic network consists of about 225 sampling stations. Some of these are in operation every year, about 80 being situated in large cities and 35 at non-urban sites. The remaining stations, divided into two groups, sample during alternate years. Thus about 175 stations are active

160

in the network in any given year. Samples of suspended particulate pollutants are taken from the air every two weeks at each operating station of the network. Several common gaseous pollutants are sampled at more than 50 stations of the network.

National Coal Board (N.C.B.). A body created by the Coal Mines Nationalization Act of 1946 to take control of some 950 coal mines in England, Wales, and Scotland. The Board's statutory duty is to develop the industry efficiently and provide coal in such quantities and at such prices as to further the public interest. Today production has been confined to the 400 or so most efficient mines. During the year ended 31st March, 1966, the N.C.B. collieries produced 174·1 million tons of deep-mined coal and 7·1 million tons of opencast coal. Licensed private mines produced 1·6 million tons. The total of 182·8 million tons was 9·7 million tons less than in the previous year. With the increasing use of *natural gas*, q.v. from the North Sea, the break-through in the economics of nuclear power, and the rising consumption of oil refinery products, the production of the N.C.B. is likely to fall dramatically in the coming years. Of the total deep-mined output over 80 per cent is mechanically cut and loaded, by far the highest degree of mechanization in Europe.

National Coal Board Coal Classification. A classification system for coals prepared by the *National Coal Board*, q.v., of the United Kingdom. The classification is based on the *volatile matter* content of coals on a *dry*, *mineral-matter-free*, qq.v., basis, and the caking power of " clean coal ". A clean coal is defined as that with not more than 10 per cent ash. There are four main groups: 100, 200, 300, and 400 to 900. Each group is divided into classes and sub-classes. Coals of groups 100 and 200 are classified by using the parameter of volatile matter alone. Those in group 100 have a volatile matter content of under 9·1 per cent and consist of anthracites. Those in group 200 have a volatile matter content of 9·1 to 19·5 per cent. In this group fall the low-volatile steam coals, dry steam coals and coking steam coals. In group 300 the volatile matter content ranges from 19·6 to 32·0 per cent. In this group fall the medium-volatile coals and prime coking coals. In these first three groups (100 to 300) there is in general a close relationship between volatile matter content and caking properties. In the fourth main group (400 to 900) with over 32 per cent volatile matter content, is a wide range of caking properties at any given volatile matter content, and sub-divisions are made on the basis of the *Gray-King Assay*, q.v. See **Coal Classification Systems.**

National Industrial Fuel Efficiency Service (N.I.F.E.S.)

National Industrial Fuel Efficiency Service (N.I.F.E.S.). Formed in May, 1954, a non-profit British organization which offers industry, local authorities and commercial property owners practical help in the use of heat and power, whatever type of fuel be used. N.I.F.E.S. provides heat and power surveys which show by precise measurement where waste occurs in factory premises and how it can be prevented; boiler house efficiency tests; advice on improving the efficiency of heat-using process plants, furnaces and power plants; checks on space-heating systems and advice on insulation; advice on the requirements of the Clean Air Act relating to smoke and grit emissions and other services.

National Survey of Air Pollution, British. An air pollution sampling network established in Britain in the early 1960's. Measurements of air pollution had been made since 1914 by Local Authorities and other interested bodies, in cooperation with the Department of Scientific and Industrial Research (now the Ministry of Technology). The original system of measurements, while providing a rough indication of local conditions was not, however, sufficiently comprehensive to provide an accurate picture of the distribution of air pollution in different types of area throughout the country. Since information could only be obtained from a survey planned on proper statistical lines and, as the Clean Air Act became fully operative and more local authorities showed interest, the need for a more systematic national approach sharpened. As a result, a Working Party consisting of representatives of local authorities, the Medical Research Council, the Meteorological Office, the Ministry of Housing and Local Government and DSIR was set up to devise a new survey. The recommended scheme was accepted by the Standing Conference of Cooperating Bodies for the Investigation of Atmospheric Pollution at their meeting on 14th November, 1960. It was recognized that the pollution caused by dust was very localized and that it would be inappropriate to include dust deposition measurements in the new survey. Although the Ministry of Technology continues to publish the results of the many deposit gauges installed throughout the country, the National Survey itself is confined to measurements of SO_2 and smoke. It was recommended that SO_2 in the atmosphere should be measured by the daily volumetric hydrogen peroxide method and smoke by means of a filter. Both were conveniently combined in the same piece of apparatus developed by DSIR. See **Volumetric SO_2 Apparatus.**

Natural Circulation. Circulation due to thermal and density effects, in contrast to forced circulation. The heating of water in a boiler results in some of the water becoming hotter than the rest; the hotter part being less dense rises, cooler water taking its place.

Natural Coal Dust. The fraction of raw coal which passes a test sieve of 60 B.S. mesh (0·25 mm aperture); it differs fundamentally from the other coal constituents in that it consists largely of fusain concentrated in the dust during mining and subsequent operations—the volatile content is low, it is feebly caking or non-caking, and the ash content lies between 12 and 20 per cent. The majority of run-of-mine coals contain from 2 to 6 per cent of natural dust; it may rise to 15 per cent in untreated smalls. This dust tends to clog a fuel bed and when dislodged by raking or forced draught aggravates the grit emission problem.

Natural Draught. Draught produced by a chimney and the relative densities of the flue gases and the ambient air, without the assistance of fan power. If the draught requirements are known, the chimney height may be calculated from the following formula:

$$H = D \left/ \left(\frac{7 \cdot 6}{t_a + 460} - \frac{7 \cdot 9}{t_m + 460} \right) \right.$$

where, H = height of chimney in feet above grate level
D = draught in in. w.g.
t_a = ambient air temperature (°F)
t_m = mean gas temperature in chimney (°F)
7·6 = wt. of 100 ft³ air at 60°F lb
7·9 = wt. of 100 ft³ flue gases at 60° F lb

Likewise, if the chimney height is given, the theoretical draught intensity will be:

$$D = H \left(\frac{7 \cdot 6}{t_a + 460} - \frac{7 \cdot 9}{t_m + 460} \right)$$

The mean flue gas temperature, if not known, may be calculated in an approximate manner by deducting 2° F from that at the base of the chimney for every 3 ft of chimney height. See **Draught.**

Natural Gas. Gas obtained from underground sources, often in association with petroleum deposits. It generally contains a high percentage of methane, CH_4, with varying amounts of ethane, C_2H_6, and inert gases such as carbon dioxide, nitrogen and helium. Most of the natural gas delivered by U.S. utilities contains less than

10 per cent of inerts. The calorific value is of the order of 1000 Btu/ft^3 (555 Chu/ft^3). See **Methane**.

Natural Gasoline. *Gasoline*, which accompanies "wet" *natural gas*, qq.v., in many regions; it is removed from the gas by compression or absorption, or both. The gasoline is "low boiling" consisting mainly of hexanes, heptanes, octanes and pentanes. The average yield of gasoline is about 1·5 gallons per 1000 ft^3 of gas treated. Also known as *casinghead gasoline*, q.v.

Natural Gas Pipelines. Distribution mains carrying natural gas from its source to its market. In Britain, a pipeline was constructed to carry imported natural gas from Canvey Island, Essex, to a terminal near Leeds, with suitable branches; the system was some 325 miles in length. Since the discovery of North Sea gas further networks are being established to serve all the principal areas of Britain. Pipelines serve most of the industrial areas of the United States, utilizing the vast natural gas resources of that country. During 1966, a 450 mile pipeline between the Tyumen gas field in western Siberia and the industrial complex in the Urals was completed. The discovery of the world's second largest natural gas field in the Netherlands in 1959 has led to extensive pipeline developments in Europe.

Natural Pollutants. Substances of natural origin present in the earth's atmosphere which may, when present in excess, be regarded as air pollutants. These include: (a) ozone, formed photochemically and by electrical discharge; (b) sodium chloride, or sea salt; (c) nitrogen dioxide, formed by electrical discharge in the atmosphere; (d) dust and gases of volcanic origin; (e) soil dust from dust storms; (f) bacteria, spores and pollens; (g) products of forest fires. See **Air Pollution**.

Neon. A rare gas. When an electric current of sufficient intensity is passed through a glass tube containing neon gas under atmospheric pressure, a red glow is produced. Other rare gases produce different colours.

Neptunium. Np. An *element*, q.v., at. no. 93. It is produced in a *nuclear reactor* by the absorption of a *neutron*, qq.v., by $_{92}U^{238}$ to give $_{92}U^{239}$; the latter subsequently emits a beta particle becoming $_{93}Np^{239}$. In turn this disintegrates into plutonium, $_{94}Pu^{239}$ by the emission of a further beta particle.

Net Calorific Value. The number of British thermal units (or Centigrade heat units) liberated when one pound of coal or oil, or a cubic foot of gas, is completely burned in oxygen saturated with water vapour, the final products being carbon dioxide, sulphur dioxide, nitrogen and water vapour. *Latent heat*, q.v., is lost in practice in the water vapour and consequently the effective heat potential of a given quantity of fuel is less than its *gross calorific value*, q.v., might suggest. A method recommended by the Institution of Civil Engineers for making a correction to the gross calorific value of a solid fuel to obtain the net or lower calorific value (in Btu) is:

$$\text{Net c.v.} = \text{gross c.v. of air dried fuel} - 1055 \left(\frac{M\% + 9H\%}{100} \right)$$

where M and H are the percentages of moisture and hydrogen respectively in the air-dried fuel. See **Calorific Value.**

Neutral Atmosphere. In meteorology, a term usually applied to conditions in the lowest layers of the atmosphere when the *lapse rate*, q.v. lies between zero and $5 \cdot 4° \text{ F } (3° \text{ C})$ per 1000 ft. See **Inversion; Superadiabatic Lapse Rate.**

Neutral Flame. In welding, a *flame*, q.v., produced by an acetylene torch from a mixture of *acetylene*, q.v., and oxygen in equal volumes.

Neutron. An uncharged particle of mass $1 \cdot 675 \times 10^{-27}$ kg, which is a constituent of all nuclei except hydrogen, $_1H^1$. Outside a *nucleus*, q.v., a neutron is radioactive decaying with a half-life of about 12 minutes into a *proton* and an *electron*, qq.v. It is the agent which promotes the fission of U^{235} in a *nuclear reactor*, q.v.

Neutron Source. Any substance that emits neutrons; for example, a mixture of radium and beryllium emits neutrons by a nuclear reaction between the beryllium nuclei and the alpha particles emitted by the radium.

Newton. Unit of force; that force which, applied to a mass of 1 kg, gives it an acceleration of 1 m/s^2.

Niobium. Nb. A rare metal and *element*, q.v.; at. no. 41, at. wt. $92 \cdot 906$, possessing a low neutron-capture cross-section; it has been used as a canning material for nuclear fuel elements operating at high temperatures, e.g. when the *coolant*, q.v., is liquid sodium.

Nitrides. Compounds of metal and nitrogen. See **Nitriding.**

Nitriding. A case hardening *heat treatment*, q.v., process for producing a hard surface on certain types of steel by heating in gaseous ammonia. The ammonia cracks at the metal surface causing the formation of *nitrides*, q.v.

Nitrogen. N. A non-metallic *element*, q.v.; atomic number 7,

Nitrogen

atomic weight 14; an odourless, colourless and, in the general atmosphere, a chemically inert gas. The predominant constituent in air it serves only as a diluent in combustion processes. See **Air.**

Nitrogen Oxides. Oxides which include nitric oxide, NO; nitrogen dioxide, NO_2; and nitrous oxide, N_2O. They are produced in the combustion of organic matter and are thus introduced into the atmosphere from automobile exhausts, furnace chimneys, incinerators and other similar sources. Oxides of nitrogen undergo many reactions in the atmosphere. Nitric oxide reacts with oxygen to form nitrogen dioxide, although there is some controversy as to the extent and speed of this oxidation process. When NO and NO_2 are present the following reaction proceeds rapidly:

$$NO + NO_2 + H_2O \rightarrow 2HNO_2 \text{ (nitrous acid)}$$

Nitrogen dioxide reacts with vapour and oxygen to give nitric acid vapour:

$$H_2O + 2NO_2 + \tfrac{1}{2}O_2 \rightarrow 2HNO_3 \text{ (nitric acid)}$$

Nitrogen dioxide can be injurious to health when inhaled; nitric acid being produced on the lung tissue with the moisture in the lungs. In the U.S. Cleveland Clinic fire in 1929 many deaths were due to NO_2 from burning X-ray film. Nitric oxide can combine with the haemoglobulin of the blood to form an addition complex. Oxides of nitrogen from motor vehicle exhausts are important in the atmosphere of Los Angeles, California, where in the presence of sunlight they catalyse the formation of ozone.

Non-caking coal. A coal which does not "cake" when heated leaving, after the *volatile matter*, q.v., has been driven off, a dustlike coke. See **Caking Coal.**

Non-ferrous Metals. Metals other than iron. See **Brass; Gunmetal; Phosphor Bronze.**

Normal Temperature and Pressure (N.T.P.). A conventional reference condition of 0° C and 760 mmHg. It is synonymous with "standard temperature and pressure" or s.t.p.

Nozzle-mix Burner. A *gas burner*, q.v., in which all the combustion air is provided by high velocity mixing at the base of the flame. Also known as a high-intensity or high-velocity burner.

Nuclear Fuel. Substances capable of producing heat as the result of the splitting or fission of their atomic nuclei, and not through the chemical process of *combustion*, q.v. Fissionable *isotopes*, q.v., are uranium 235 (from ore), uranium 233 (bred in reactor) and plutonium 239 (bred in reactor). A pound of uranium 235 is capable of producing $3\cdot3 \times 10^{10}$ Btu, or approximately 2,750,000 times the amount of heat produced from burning a pound of coal.

Nuclear Reactor. A device in which nuclear *fission*, q.v., takes place as a self-supporting chain reaction. A typical reactor comprises: (a) fissile material such as uranium or plutonium; (b) *moderator*, q.v.; (c) reflector; (d) control elements; (e) provision for the removal of heat by means of a *coolant*, q.v. See **Advanced Gas-Cooled Reactor (AGR); Boiling Water Reactor; Converter Reactor; Dounreay Fast Reactor; Dragon High Temperature Reactor; Dungeness 'B' Nuclear Power Station; Heterogeneous Reactor; Homogeneous Reactor; Magnox Reactor; Pressurized Water Reactor; Pressurized Water Thorium-Uranium Converter Reactor; Savannah, N.S.; Swimming Pool Reactor; Zero Energy Reactor.**

Nucleon. A constituent of the atomic *nucleus*; a *proton*, or a *neutron*, qq.v.

Nucleus. The positively-charged core of an *atom*, q.v., which contains practically the whole mass of the atom while possessing only a minute fraction of its volume. It has a diameter of between 10^{-12} and 10^{-13} cm, and contains neutrons and protons. The charge equals the *atomic number*, q.v.

Nuclide. A type of atom characterized by a given number of protons and neutrons in its nucleus.

O

Oak Ridge Formula. A formula for calculating the ground level concentration of a pollutant emitted from an elevated source:

$$C = \frac{315 \times 10^6 W}{h(14vd + Q)} \text{ p.p.h.m.}$$

where: C = maximum ground level concentration of SO_2
          p.p.h.m. (parts per hundred million by volume)
   W = strength of source—lb/s. of SO_2
   v = velocity of efflux—ft/s.
   d = stack diameter—ft
   h = stack height—ft
   Q = rate of heat output relative to air temperature—Btu/s.

This formula has been deprecated because of its empirical nature. However, Best ("Maximum Gas Concentrations at Ground Level from Industrial Chimneys" *J. Inst. Fuel* 1957, **30**, 329) has checked the use of the Oak Ridge Formula to calculate ground level concentrations and concluded that it gives results which are in satisfactory agreement with figures calculated using the work or either Sutton or Bosanquet. Consequently, the above formula which is easy to use is regarded as useful for the estimation of maximum ground level concentrations of SO_2. With normal wind distribution patterns the long-term concentrations will be very much lower.

Oberhausen Rotor Furnace. A *steelmaking furnace*, q.v., of German origin, comprising a very slowly rotating barrel-shaped vessel. A primary oxygen lance projects oxygen into the bath of molten pig iron, while a secondary lance blows oxygen over the bath. This secondary oxygen supply burns the carbon monoxide released from the boiling metal.

O.C.C.R. Gasifier. A gasification unit in which fuel oil is gasified; partial combustion and cracking occurs on the base of a refractory lined vertical chamber. The product gas then passes to the furnace where secondary air is admitted to complete combustion. The calorific value of the hot raw gas is about 200 Btu/ft³ (110 Chu/ft³).

Octane Number. A method of ranking gasolines according to their resistance to detonating explosions when used as fuels in internal combustion engines. An octane number is numerically equal to the percentage by volume of iso-octane (2,2,4-trimethyl pentane) in a mixture of normal heptane and iso-octane having the same knocking tendency as the fuel being tested. This mixture is used because n-heptane has bad anti-knock properties and iso-octane has extremely good anti-knock properties. The scale is based on 0 for n-heptane and 100 for iso-octane. Gasolines may also be rated above 100 octane by use of extrapolation formulae. Octane number is the universal scale used to define the anti-knock or anti-pinking qualities of a motor spirit. See **Aviation Mixture Methods; Gasoline Additives; Motor Method; Research Method; TetraEthyl Lead.**

Octopus System. A system of firing pulverized coal; the airborne fuel is distributed to several firing points by supply lines from a single source. It is in use, for example, for top-fired annular brick kilns, the coal being delivered under control to several rows of feedholes on the kilns so as to give a curtain of flame across the kiln at the different rows.

Odorizer. A chemical substance added to a gas to give it a distinctive smell. *Town gas*, q.v., produced from oil feedstocks requires an odorizer to be added as a statutory safety precaution. The most commonly used substance is tetrahydrothiophene.

Ohm. The unit of electrical resistance; the electrical resistance between two points of a conductor when a constant difference of potential of 1 volt, applied between these two points, produces in the conductor a current of 1 *ampere*, q.v., this conductor not being the source of any electromotive force. This fundamental relationship between voltage and current is known as Ohm's Law; named after G. S. Ohm, a German physicist (1787–1854).

Oil. See **Petroleum.**

Oil Burner. A piece of equipment whose function is to deliver fuel into a *combustion chamber*, q.v., in a form suitable for combustion, this is achieved by vaporizing or "atomizing" the fuel. See **Air Assisted Pressure Jet Burner; Blast Atomizer; Gun-type Burner; Inside Mix Burner; Outside Mix Burner; Pressure Jet Burner; Rotary Burner; Self-proportioning Burner; Spill Type Burner; Steam Blast Burner; Vaporizing Burner; Wide Range Pressure Jet Burner.**

Oil Gas. Gas manufactured by the gasification of oil with steam in a chamber containing hot chequer-brickwork; the process is a cyclic one as the reactions are endothermic. See **Endothermic Reaction.**

Oil Preheating. See **Primary Heating; Secondary Heating.**

Oil Refining. The process by which crude oil is divided up and made into marketable products.

Oil Slug Injection System. A system for firing top-fired continuous brick kilns; oil is circulated through a *ring main*, q.v., each separate row of feed holes being supplied by a branch line connected between the main flow and return lines. The branch line incorporates off-take points at each feed hole; a regulating valve at each feed hole controls the size of the oil slug. At the end of each branch line a control panel regulates the number of slug injections, normally ranging from 1 to 12/min.

Oil Synthesis. A process for the production of oil from hydrogen and carbon monoxide.

Oil Traps. Geological configurations which permit the trapping of oil. The two main types of trap are structural and stratigraphic. Structural traps include domal traps, in which the rock beds have been formed into an inverted dome; the anticlinal trap, a kind of elongated dome; the fault trap, formed by a fracture plane running down through layers of rock making it possible for the sediments on each side of the fracture to slip out of alignment with an impervious layer being brought opposite a tilted reservoir rock. Stratigraphic traps can be formed in several ways. One example is that of the up-dip edge of a wedging-out sand layer grading into an impermeable clay or shale. Another type of stratigraphic trap is sometimes provided by an ancient reef buried by impervious sediments. An oil reservoir is not an underground lake; it is an accumulation of liquid hydrocarbons within the pores of a particular kind of rock.

Olefin Series. C_nH_{2n}. Aliphatic *hydrocarbons*, q.v., in which two carbon atoms may share more than one bond; these compounds are described as "unsaturated", since by breaking one of the double links additional atoms of hydrogen or other elements may be added. They can also combine with each other by forming cross-linkages; that is, they polymerize. Examples of olefins are: (a) ethylene, C_2H_4; (b) proplylene, C_3H_6; (c) 1 butene, C_4H_8. This type of hydrocarbon is formed when oil is subjected to high temperatures in cracking processes.

Oleum Spirits. A petroleum "cut" between 300 and 400° F (149 and 204° C) meeting certain other specifications.

On-load Refuelling. The replacement of the fuel elements of a *nuclear reactor*, q.v., while the reactor is on-load.

On-off Control. An automatic device by which the heat being supplied to a plant is cut off when the temperature exceeds a set point; in this system heat is supplied at a prescribed rate, or not at all. See **Proportional Response Control.**

Onia-Gegi Process. A cyclic gasification process based originally on heavy oil and later adapted to accept light distillates. The process is one of partly thermal and partly catalytic cracking in the presence of an excess of steam and a nickel-bearing catalyst, with a process temperature of between 1400 to 1472° F (760 to 800° C). The calorific value of the gas produced is about 500 Btu/ft^3.

Open Cut Mining. Or strip mining, a technique of mining employed when the coal is not more than about 100 ft below the surface; the overlying earth and rock are mechanically stripped to expose the coal which is then removed with or without blasting.

Open Cycle. A mode of operation of a *heat engine*, q.v., in which the working fluid is used only once.

Open Flash Point. A *flash point*, q.v., determined in an "open" apparatus, the sampling cup having no cover. The open flash point of a liquid is a few degrees higher than the closed flash point. See **Abel Flash Point Apparatus; Pensky-Martens Flash Point Apparatus.**

Open-Hearth Furnace. A *steelmaking furnace*, q.v., consisting of a shallow bath capable of holding 60 to 400 tons of metal at a time; the smaller furnaces are static, the larger tilting. Construction is of special bricks inside a steel casing. In the end wall are situated the oil or creosote-pitch burners and ports for the admission of pre-heated combustion air; in earlier designs using coke-oven or pro-ducer gas, provision was made for this to be preheated also. The preheating of air (and fuel gas, if used) takes place in regenerative heat exchangers situated below the furnace; waste gases from the furnace flow through the regenerators before passing to a *waste-heat boiler*, q.v., and hence to the chimney. The flow of the gases is reversed every twenty minutes or so, the combustion air flowing alternately through the regenerators at each end of the furnace. A typical charge consists of 60 per cent steel scrap and 40 per cent pig iron; the three stages of charging, melting and refining take about 10 to 14 hours.

Open-loop Control System. A control system in which a controlled variable is allowed to vary in accordance with the inherent charac-teristics of the control system and the controlled power apparatus for any given adjustment of the controller. See **Closed-Loop Control System; Feedback Controller.**

Opencast Mining. The working of coal seams near their outcrops, i.e. near the point at which coal appears at the surface. Generally, the actual outcrop edge of a seam is very inferior, due to prolonged oxidation, and has to be discarded.

Optical Pyrometer. A device for measuring temperatures, the hot object being observed through a telescope inside which a standard-ized electric lamp is fitted. The current passing through the lamp is adjusted until the brightness of the filament matches the brightness of the object. The current is measured with an ammeter whose scale is calibrated in units of temperature. A colour match may be made against a standardized tungsten filament. The instrument is suitable for temperatures ranging from about 1290° F (700° C) up to about 6330° F (3500° C). See **Radiation Pyrometer.**

Orifice Meter. A flowmeter employing as a measure of flow the pressure differential across an orifice, i.e. the pressure measured on the upstream and downsteam sides of the orifice, as fitted in a pipe or duct.

Orsat Apparatus. Instrument for the determination of *carbon dioxide*, *oxygen* and *carbon monoxide*, qq.v., in flue gases. It con-sists essentially of three glass vessels which are partly filled with

171

absorbent liquids through which a measured sample of flue gas may be passed. A solution of potassium hydroxide (caustic potash) is used to absorb carbon dioxide; a mixture of pyrogallic acid in caustic potash is used to absorb oxygen; finally a solution of ammoniacal cuprous chloride is used to absorb carbon monoxide.

Orthohydrous Coal. *Coal*, q.v., of normal or typical *hydrogen*, q.v., content for the type species. See **Per-hydrous Coal; Sub-hydrous Coal.**

Ostwald Diagram. A graph showing the relationship between carbon dioxide and oxygen for a particular fuel; the oxygen in the air is graphically connected to the maximum possible carbon dioxide in the flue gases for the fuel concerned. The oxygen and carbon dioxide readings should fall along this line. See Fig. 18.

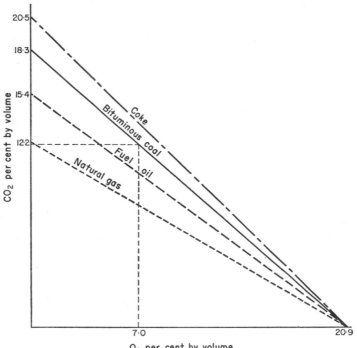

FIG. 18. *Ostwald diagram.*

Ostwald Viscometer. A *viscometer*, q.v., for measuring the *viscosity*, q.v., of kerosine, gas oils and diesel oils. The viscosities are expressed in *centistokes*, q.v.

Outage. The amount by which plant availability differs from the total capacity of the system through its being out of service due to breakdown, essential maintenance, or other causes.

Outcrop Coal. Part of a coal seam visible at the surface of the ground.

Outlet Flow Heater. A heater situated at the outlet of a liquid fuel storage tank; it has an immediate effect on the outlet temperature of the oil.

Outside Mix Burner. *Oil burner*, q.v., in which the oil is released into the atomizing fluid at the outlet from the burner.

Over-deck Superheater. See **Superheater.**

Overfeed Combustion. *Combustion* in which *ignition*, q.v., takes place from the bottom of the charge of fuel, the *ignition plane*, q.v.,

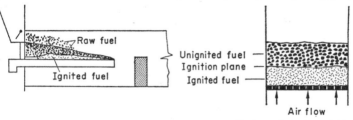

FIG. 19. *Overfeed combustion as illustrated by the sprinkler stoker.*

travelling upwards in the same direction as the air flow. The principle is illustrated in Fig. 19. This type of combustion tends to smoke formation, and if the coal has any caking properties these are well developed during combustion. See **Underfeed Combustion.**

Overfire Jets. Jets of air or steam directed over a fire bed in order to promote turbulent mixing, thus helping to achieve smokeless combustion. There are three types of jet: (a) secondary air jet (separate fan); (b) steam induced secondary air jets; (c) straight steam jet. Jets must be so designed as to penetrate the depth of the furnace, and sufficient in number to ensure turbulence over the whole area of the fuel bed.

Overhead. In an oil refinery distilling operation, that portion of the charge which is vaporized.

173

Oxidant. In the study of *air pollution*, q.v., any chemical substance with an oxidation potential greater than that of oxygen, e.g. ozone, O_3.

Oxidation. The addition of oxygen to a compound; the reverse of *reduction*, q.v. An oxidizing atmosphere is one containing free oxygen.

Oxide. An ash formed by the union of oxygen with, for example, a metal. The rusting of iron is the result of oxidation, the rust being described as iron oxide.

Oxides of Nitrogen, See **Nitrogen Oxides.**

Oxy-fuel Burner. A *burner*, q.v., in which a supply of gaseous or liquid fuel and a supply of oxygen are delivered to the same nozzle. The fuel may be natural or other gas, or oil. Oxy-fuel burners have been employed in the steel industry as an alternative to oxygen lancing, producing the same result without copious emissions of iron oxide fume. The simplest oxy-gas burner consists of two concentric tubes with oxygen supplied through the central tube and gas through the annulus. Burners usually incorporate an arrangement to produce sufficient turbulence to mix thoroughly the oxygen and fuel.

Oxygen. A non-metallic *element*, q.v.; symbol O, atomic number 8, atomic weight 16; an odourless, colourless gas which supports both combustion and life. Commercial or tonnage oxygen is produced from liquid air by fractional distillation. Oxygen as such is used in oxy-acetylene torches, oxy-fuel burners and in oxygen lances for steel-making. See **Air.**

Oxygen Recorder. An instrument which uses the magnetic properties of oxygen for measurement purposes; oxygen is paramagnetic while the other principal components of flue gases are diamagnetic. In one instrument a Wheatstone Bridge circuit is utilized, the "out-of-balance" in the circuit being proportional to the oxygen present. See **Orsat Apparatus.**

Ozone. O_3. A gaseous oxidizing agent which has been used in air conditioning; above minimal levels, however, it is an irritant to human beings and animals. It is a natural constituent of the atmosphere occurring in concentrations of about 0·01 ppm; the toxicity threshold for workers is 0·1 ppm. Ozone is produced by certain high voltage electrical equipment; it is also produced in certain circumstances when photochemical reactions occur in the atmosphere between ultraviolet light (sunlight) and the oxides of nitrogen and hydrocarbons emitted to the atmosphere by motor vehicles. See **Los Angeles Smog; Nitrogen Oxides; Oxidant.**

Package Boiler. A compact steam generator supplied as a self-contained unit complete with all fittings, draught equipment, pump, controls, etc., the whole being mounted on a bedplate. This type of boiler can be installed speedily, connections to fuel supply, water and electricity, being all that is necessary. Package boilers operate up to 400 lb/in² pressure and evaporative capacities range up to 40,000 lb steam/h. Thermal efficiency is in excess of 80 per cent. Strict attention must be paid to the quality of the feed water. See **Boiler.**

Pair Production. The conversion of a gamma-ray *photon* into an *electron*, and a *positron*, qq.v.

Paraffin Series. CnH_{2n+2}. Aliphatic *hydrocarbons*, q.v., in which the carbon atom has four bonds, to which other atoms may be attached; these compounds are described as "saturated" because each atom of carbon is singly linked to four atoms of hydrogen, the maximum number with which carbon can combine. Examples of paraffins are: (a) methane, CH_4; (b) ethane, C_2H_6; (c) propane, C_3H_8; (d) n-butane, C_4H_{10}; (e) iso-butane, C_4H_{10}.

Paraffin Wax. Wax-like substances obtained as a residue from the distillation of petroleum; m.p. in range 113 to 149° F (45 to 65° C); sp. gr. 0·9.

Parent. A radionuclide which disintegrates to yield a given nuclide, known as "the daughter", either directly or as a later member of a radioactive series. See **Daughter.**

Parr Classification. A system, proposed by S. W. Parr, for the classification of coal by rank in the natural series from anthracite to lignite. It is based upon the heat value of the true coal substance, free from ash and sulphur (i.e. mineral-matter-free), according to fixed carbon on the dry basis, or British thermal units on the moist basis. The Parr formula for the latter is:

$$\text{Btu (m.m.f. moist basis)} = \frac{\text{Btu} - 50\cdot5}{100 - (1\cdot08\,A + 0\cdot55S)} \times 100$$

where S and A are the percentages of sulphur and ash respectively. See **Coal Classification Systems.**

Partial Combustion. *Combustion*, q.v., in which the supply of air (or oxygen) is so restricted that it is incomplete. For example, in the process for producing synthesis gas from methane by reaction with a limited amount of oxygen, the products are carbon monoxide and

Partial Combustion

hydrogen; with sufficient oxygen to complete combustion the products would be simply carbon dioxide and water:

$$CH_4 + \tfrac{1}{2}O_2 \rightarrow CO + 2H_2$$
$$CH_4 + 2O_2 \rightarrow CO_2 + 2H_2O$$

Partial Combustion Reactor. See **Reactor.**

Partial Pressure. In connection with a mixture of gases, the pressure of one of the components of the mixture.

Particle Size Distribution. The percentage by weight or number of each of the specified fractions into which a sample of particulate matter is divided.

Partition Curve. A curve which indicates for each *specific gravity*, q.v., or size fraction, the percentage of it contained in one of the products of the separation. See **Ecart Probable.**

Pascal. Unit of pressure; the pressure produced by a force of one newton applied, uniformly distributed, over an area of one square metre. See **Newton.**

Pass-out Turbine. A *steam turbine* q.v., from which steam is bled for process work or heating; the remaining steam passes through the low pressure zone of the turbine to the condenser. See **Back-pressure Turbine.**

Peak Load. A transient maximum demand on a source of supply such as a steam generator, gas or electricity undertaking. Peak loads tend to occur regularly at certain hours of each day; they also occur at certain times of the week and characterize certain seasons.

Peat. The youngest member of the coal series, a *fuel*, q.v., consisting of layers of dead vegetation in varying degrees of decomposition occurring in swampy hollows in cold and temperate regions. Fresh plant growth at the surface adds material to the decomposing debris; peat may be found in layers many feet thick. Light in colour near the surface, at deeper levels it is brown and even black. The amount of moisture in peat ranges up to 90 per cent, but more typically 25 to 45 per cent. It is a bulky fuel and even when well-dried its *calorific value*, q.v., is only about 7000 Btu/lb. Its ash and sulphur content is very low. Extensively used in Scotland and Ireland it is dug out and dried during the summer months. The Electricity Supply Board of Ireland has been using peat for electricity generation for the past 15 years; total peat fired generating capacity is now 407 MW. Peat is available in two forms: (a) sod peat which consists mainly of large irregularly shaped sods of typical dimensions 10 in. × 3 in. × 3 in.; (b) milled peat which consists of multisized fibrous particles of which about 75 per cent pass through a $\frac{1}{4}$ in. sieve. Sod peat is burned in boilers with chain grate stokers and a maximum capacity of 20 MW; milled peat is fired as pulverized fuel in boilers of up to 40 MW capacity. For the same heat content, sod peat has four times, and milled peat eight times, the bulk of normal coal.

Pebble Stove. A *heat exchanger*, q.v., for providing preheated air or other gas for use in industrial processes. In its simplest form the stove consists of small refractory pebbles enclosed in a brick-lined steel shell, thus offering an alternative to conventional *chequer-brickwork*, q.v., arrangements. The initial heating of the bed may be achieved by the combustion of fuel or the utilization of waste gases; after a sufficient degree of heating has been achieved, the hot bed is available for the heating of cold air or other gas. In this form the process is cyclic and more than one stove is necessary to achieve a constant supply of preheated air. Two chamber arrangements are available in which the pebbles are heated in an upper chamber and then pass through to the lower chamber where they heat the incoming air; after cooling the pebbles are returned to the first chamber. The process being continuous, one such unit only is required to supply a continuous stream of preheated air. See **Blast-furnace Stove; Howden-Ljungstrom Air Preheater.**

Penetration Test. A test to determine the degree of hardness of bituminous material; penetration is expressed as the distance that a standard needle vertically penetrates a sample of the material under prescribed conditions.

Pensky-Martens Flash Point Apparatus. Apparatus for determining the *flash point*, q.v., of liquid fuels with flash points above 120° F (49° C). A sample of oil is heated in a closed vessel until a temperature is reached at which the vapours in the air space above are sufficient to form an inflammable mixture and ignite when a flame is applied. The flash point so determined is known as the "closed" flash point; if the test is carried out with the sample cup uncovered an "*open*" *flash point*, q.v., is determined. See **Abel Flash Point Apparatus.**

Pentanes. C_5H_{12}. Low boiling paraffin hydrocarbons. The n-pentane has a b.p. of 100° F (37·8° C) and a sp.gr. of 0·63. See **Liquefied Petroleum Gases.**

Percolation. A situation in which a fuel/air mixture becomes over rich due to the evolution of vapour in a *carburettor*, q.v., which causes fuel to percolate into the inlet manifold of the engine.

Perfect Gas. A gas which, at all temperatures and pressures, satisfies the relationship

$$pv = RT$$

where: p = pressure

v = volume occupied by one mol of the gas

R = gas constant

T = absolute temperature.

The gases hydrogen, helium, oxygen and nitrogen give the nearest approach to a " perfect gas ". Also known as an " ideal gas ".

Performance Number. The percentage gain in "knock limited power" developed by a typical supercharged aviation gasoline engine when operating on leaded iso-octane, compared with that obtained when operating on unleaded iso-octane.

Per-hydrous Coal. *Coal*, containing more *hydrogen*, qq.v., than is normal for the type species. See **Orthohydrous Coal; Sub-hydrous Coal.**

Petrochemicals. Chemicals manufactured from the products of oil refineries; mainly a post-war development based largely on ethylene, propylene and butylene produced in the cracking of gasoline fractions. It is the unsaturated olefine compounds which are used to produce chemicals; the saturated paraffin compounds (methane, ethane, propane and butane) are used as refinery fuel or sold as bottled gas or supplied for *town gas*, q.v., enrichment.

Petrol. See **Gasoline.**

Petrol Engine. See **Gasoline Engine.**

Petroleum. According to generally accepted theory, a substance derived from the remains of plant and marine life that existed on the earth millions, in some cases hundreds of millions, of years ago. A complex and variable substance, petroleum ranges from solid bitumen, through liquid oils, to highly volatile natural gases such as *methane*, q.v.; all are essentially mixtures of compounds made up from the elements *hydrogen*, and *carbon*, qq.v., being known as "natural hydrocarbons". See **Crude Oil; Hydrocarbons; Petroleum Products.**

Petroleum Coke. A residue remaining after the complete distillation of oil; the two principal types are known as "delayed process coke" and "fluid process coke", both being produced in a fine granular form. They are usually high in sulphur content, up to 10 per cent. See **Delayed-coking Process; Fluid Coke Process.**

Petroleum Products. Refinery products which include motor spirits, aviation turbine fuels, kerosines, diesel oils, furnace oils, liquefied petroleum gas, bitumen, lubricating oils and greases, and all other products derived directly or indirectly from crude oil through a refining process.

Petrology. A branch of geology concerned with the study of the individual mineral components, structure and history, of rock masses, including coal.

pH. A term used to express the degree of acidity or alkalinity of a solution; a pH value is the logarithm, to the base 10, of the reciprocal of the concentration of hydrogen ions in an aqueous solution. Thus, hydrogen ion concentration $= 1 \times 10^{-x}$ where x equals the pH. A neutral solution contains 1×10^{-7} gramme equivalents of hydrogen ions per litre; the pH value is therefore 7. Acid solutions contain more than 1×10^{-7} gramme equivalents of hydrogen ion per litre and consequently the pH values are less than 7; conversely, alkaline solutions have pH values greater than 7. The range within which pH values are expressed covers the scale 0 to 14. Acidities or alkalinities greater than those represented by this scale are expressed as "per cent concentrations".

"Phimax". A *gas coke*, q.v., of a highly reactive nature produced in continuous vertical retorts from specially selected coals; it is a smokeless fuel, very free burning, and suitable for any domestic appliance. "Phimax" is only available in the North-West of England. See **Authorized Fuels; Smoke Control Area.**

Phosphor Bronze. An alloy of copper, tin and phosphorus, with or without the addition of other elements; these alloys are used where resistance to corrosion or wear is required, e.g. as in bearings or steam pipe fittings.

Photochemical Reaction. A chemical reaction which may occur when a substance is exposed to radiation, mainly visible and ultra-violet radiation. See **Los Angeles Smog; Smog.**

Photoelectric Cell. A device in which electrons are stimulated by the action of light energy to create an electromotive force. This principle is put to practical use in measuring the density of smoke in a chimney, the smoke passing between a lamp and the photoelectric cell upon which its rays have been focused. See **Smoke Density Indicator.**

Photon. A quantum of electromagnetic radiation. It has an energy hv, where h is *Planck's constant*, q.v., and v is the frequency of the radiation, or hc/λ where λ is the wavelength of the radiation and c is the velocity of light.

179

"Phurnacite". A solid smokeless fuel in the form of carbonized ovoids; the briquettes are made from fine *Welsh dry steam coal*, q.v., and pitch. The ovoids burn for long periods without attention and are suitable for use in room heaters, boilers and cookers. See **Authorized Fuels; Smoke Control Area.**

Phytotoxicant. A chemical agent that produces a toxic effect in vegetation.

Pick Breaker. A device for reducing the size of large *run-of-mine*, q.v., coal. Strong pick blades mounted rigidly on a solid steel frame move slowly up and down, the coal passing under the picks on a slowly moving horizontal plate conveyor belt. Several machines may be placed in series, with screens in between to remove fines. See **Bradford Breaker.**

Pickling. The use of hot or cold acid solutions to remove oxides and scale from metal surfaces.

Picocurie. One millionth of a *microcurie*, q.v., (10^{-12} curie).

Pile. An obsolescent term for a *nuclear reactor*, q.v., derived from the first reactor which consisted of a pile of graphite blocks containing pieces of uranium.

Pinking. See **Knocking.**

Piston Type Gas Holder. A vertical steel plate tower, with a moving piston floating on stored gas. See **Wet Gas Holder.**

Pitch. See **Coal Tar Fuels.**

Pitot Tube. A device for the measurement of fluid flow in ducts. In one form it consists of two concentric tubes, the centre one terminating at the tip of the pitot tube in a pitot head hole. The outer tube has a series of holes drilled in it, the pitot static holes. When the pitot tube is placed in a flowing gas with the nose pointing upstream the static holes, because there is no gas flowing into or out of them, measure the static pressure of the gas. The pitot head hole measures the combination of static pressure and velocity pressure due to the movement of the gas. The difference between the two pressures can be used as a measure of gas velocity. For this purpose the chambers at the back of the holes are connected to two tappings at the other end of the pitot tube. The pressure difference is measured by an inclined gauge manometer. The velocity of gas is calculated from the formula:

$$V = 58 \cdot 58 \sqrt{\frac{T}{sP}} \cdot \sqrt{h} \text{ ft/s}$$

where h = pitot pressure difference, in. w.g.

s = specific gravity of flue gas relative to dry air at absolute temperature (°R)

P = pressure, in. w.g. abs.

T = absolute temperature (°R).

Planck's Constant, h. The ratio of a quantum of radiant energy of a particular frequency to the frequency; the value is $6\cdot6253 \times 10^{-27}$ erg per second.

Plant Load Factor. See **Load Factor, Plant.**

Plant Tissues. Complex compounds comprising carbohydrates such as cellulose, lignin, starches and sugars, and proteins such as fats, oils, resins and waxes. The carbohydrates, which form a high proportion of most plant tissues, consist of carbon, hydrogen and oxygen; the proteins are compounds of carbon and hydrogen with little or no oxygen. Plants synthesize their component tissues from atmospheric carbon dioxide and mineral charged waters from the soil.

Plate Precipitator. An *electrostatic precipitator*, q.v., which comprises a number of vertical plates between which the gas passes horizontally. Plates are about 10 in. apart and from 20 to 28 ft in height; discharge electrodes hang centrally between them at about 10 to 12 in. intervals. Plate precipitators are usually subdivided into a number of zones in series which may be energized and rapped separately.

Platformer. An oil refinery reforming process employing a platinum catalyst. Naphthas vaporized by heat and mixed with recycle hydrogen are passed through the catalyst beds. The catalyst promotes molecular rearrangements among the hydrocarbons to yield a reformed product; the platformate, containing a high percentage of aromatics which have great value in the compounding of high grade *gasoline*, q.v.

Plutonium, Pu. A product of the radioactive decay of *neptunium*, q.v.; contains isotopes 238 and 239. Plutonium is an important by-product in nuclear reactions; it is recovered during the chemical processing of irradiated fuels. Used in the manufacture of weapons, it is hoped that it will ultimately be used as a fuel in fast reactors of the type being developed at Dounreay. See **Dounreay Fast Reactor; Windscale Chemical Processing Plant.**

Pneumatic Controller. An automatic control system in which a Bourdon tube is used as a temperature measuring device; the movement of the Bourdon element is communicated to a fuel controlling valve by air pressure on a diaphragm. By the introduction of a "feedback" bellows into the system a proportional response control can be achieved. See **Bourdon Gauge; Proportional Response Controller**.

Pneumatic Conveyor. A system by which loose material, e.g. fine coal, is conveyed through tubes by a stream of air. The air stream may be created by the expansion of compressed air through nozzles.

Poise. The unit of *dynamic viscosity*, q.v., in the *metric system*, q.v. Viscosities are usually tabulated in centipoise where 1 poise = 100 centipoise. See **Viscosity**.

Pole's Formula. A formula for calculating the quantity of gas flowing at low pressures through a large circular pipe, up to about 8 in. w.g. pressure:

$$Q = 1350 \sqrt{\frac{\Delta \rho d^5}{LS}}$$

where Q = gas flow, ft³/h at N.T.P.

$\Delta \rho$ = pressure drop through pipe, in. w.g.

d = diameter of pipe, in.

L = length of pipe, yd

S = specific gravity of gas (air = 1).

The length of L is increased by 5 ft for each sharp elbow or tee, and by 2 ft for each 90° bend. This modification of *Fanning's Equation*, q.v., assumes a friction factor of 0·026.

Pollutants of Natural Origin. See **Natural Pollutants**.

Polymerization. An oil refinery process for producing high octane gasoline components; it is a normal adjunct to a catalytic cracking unit. The process results in the combination of several molecules to form a more complex molecule, possessing the same *empirical formula*, q.v., as the former.

Porosity. In respect of coke, the fraction of the total volume which is represented by the pores in the pieces; it is defined:

$$1 - \frac{apparent\ specific\ gravity,\ q.v.}{true\ specific\ gravity,\ q.v.}$$

See **Voidage**.

Positive Displacement Meter. A self-contained meter for measuring feed-water flow on the delivery side of a feed pump; a type in which a piston sweeps a definite volume of water along the feed pipe at every stroke. See **Feed-water Meter.**

Positron. A particle identical to the *electron*, q.v., except that its charge is positive. It is emitted in the beta decay of many radioactive atoms and is formed in the process of *pair production*, q.v.

Post-aerated Burner. A *gas burner*, q.v., in which all the air required for combustion is introduced as secondary air.

Potential Temperature, Gradient of. In the lowest layers of the atmosphere, the difference between the adiabatic lapse rate and the actual atmospheric lapse rate. When the gradient of potential temperature is zero, the buoyancy of a mass of rising gas is constant and there is no theoretical ceiling. If the gradient is negative, the buoyancy of the gas will increase as it rises, and if it is positive it will decrease and there will be a height at which the buoyancy vanishes. See **Adiabatic Process; Lapse Rate.**

Pour Point. That temperature which is 5° F (2·8° C) above the temperature at which the oil just fails to flow when cooled under prescribed test conditions. See **Pour Point Depressant.**

Pour Point Depressant. A lubricating oil additive intended to lower the temperature at which oil will just flow; typical compounds are wax-naphthalene condensates or polymeric compounds. These depressants are only suitable for waxy (paraffinic) oils.

Power Factor. In its application to *alternating current*, q.v., a measure of the proportion of useful current employed to that expended in overcoming the effects of inductance and capacity.

Power Forming. An oil refinery unit in which low octane naphtha is converted to a high octane component of premium motor spirit; the naphtha is first hydrofined and then reformed over a platinum catalyst in an atmosphere of hydrogen at high pressure.

Power Kerosine. See **Vaporizing Oil.**

Prandtl Number. A dimensionless group of factors which serve as a criterion of temperature gradient similarity in fluids; it is expressed:

$$\frac{C_p \mu}{K}$$

where C_p = specific heat

K = thermal conductivity

μ = viscosity.

Pre-aerated Burner. A *gas burner*, q.v., in which the primary air is mixed with gas before arrival at the burner.

Pre-boiler Burner

Pre-boiler Burner. An arrangement whereby solid fuel is consumed in a separate furnace fitted to the front of the boiler, the hot gases produced in the burner passing through into the boiler proper; the furnace is circular in section, water-cooled and supplied with fuel from a hopper mounted on the top. A fan provides primary and secondary air to the furnace. It is claimed that the fitting of a pre-boiler unit to a *sectional boiler*, q.v., can result in a fuel saving of 15 to 18 per cent.

Precipitation Sampler. Automatic sampler; samples are collected in receptacles designed to open only during periods of rainfall. Thus these samples contain only the pollutants carried by precipitation and do not collect ordinary dust-fall.

Preferential Combustion. When a compound of two combustible elements burns in an insufficient amount of air, the taking of the limited amount of oxygen by the element with the greater affinity for oxygen, whicle the other element is liberated. For example, if hydrogen sulphide is burned in an insufficient supply of air, the combustion of hydrogen takes preference over the combustion of sulphur, and sulphur is deposited:

$$2H_2S + O_2 \rightarrow 2H_2O + 2S$$

This principle is adopted to achieve sulphur recovery in oil refineries.

Premix Burner. A *burner*, q.v., in which the fuel and oxidizer are mixed prior to ignition. See **Direct Burner.**

Prepared Town Gas (P.T.G.). A specially prepared gas used in carburizing processes. It is produced by the thermal decomposition of coal gas in the presence of iron turnings which removes *oxygen*, q.v., and gives a satisfactory $CO : CH_4$ ratio. The gas provides a "balanced" furnace atmosphere in which the *heat treatment*, q.v., of steel may be conducted without scaling, *carburizing* or *decarburizing*, qq.v.

Pressure Distillation. Petroleum distillation in stills under pressure to produce a cracked, light, gasoline bearing distillate product.

Pressure Gasification of Coal. See **Lurgi Process.**

Pressure Gauge. An instrument fitted to every steam boiler to indicate steam pressure. It is usually of the Bourdon type. The dial should be not less than six inches in diameter, with graduations in pounds per square inch ranging from zero to twice the operating pressure. An exception to this is the *critical pressure gauge*, q.v. The dial must indicate: (a) the boiler operating pressure in red; (b) the maximum permissible working pressure in purple. The gauge requires a stop valve so that it may be removed while the boiler is under steam. See **Bourdon Gauge.**

Pressure Jet Burner. An *oil burner*, q.v., in which oil is supplied under pressure to a special nozzle which converts the pressure energy of the oil into kinetic energy; the oil is forced through tangential slots in a sprayer plate to impart a high rotational speed to the oil, the oil then leaving the burner through a swirl chamber and precision orifice which controls the conical angle of the oil mist. See Fig. 20.

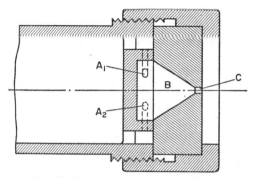

FIG. 20. *Typical pressure jet nozzle.*
A_1, A_2. *Tangential slots.*
B. *Swirl chamber.*
C. *Precision orifice.*

The mist droplets vary in size between about 20 μ and 200 μ. The oil is supplied to the burner at pressures of 250 up to 1000 lb/in². A *turn-down ratio*, q.v., of about 3 to 1 only is attainable unless special wide range pressure jets are used. See **Atomization; Wide Range Pressure Jet Burner.**

Pressure Vessel. The vessel, containing the fuel elements, *moderator* and *coolant*, of a *nuclear reactor*, qq.v. Its purpose is to enable the reactor to be operated at pressures above atmospheric in order to improve the heat transfer properties of the coolant. It is generally constructed of steel, or of prestressed concrete.

Pressurized Water Reactor. A *nuclear reactor* using water as the *moderator* and *coolant*, qq.v., in which the system is maintained at a high pressure to prevent boiling. The water is circulated through a heat exchanger to generate steam in a secondary circuit.

Pressurized Water-Thorium-Uranium Converter Reactor

Pressurized Water-Thorium-Uranium Converter Reactor. A nuclear reactor installed at Indian Point Power Station, 24 miles north of New York City on the Hudson River, U.S.A., by the Consolidated Edison Company of New York; the station has an electrical output of 275 MW. The reactor operates at 1485 lb/in² and is cooled by the circulation of water. The reactor core consists of 120 fuel elements; each of the fuel elements contains 195 fuel rods arranged in a square lattice. The rods are stainless steel tubes loaded with fuel pellets. Thorium 232 does not itself undergo fission but absorbs neutrons to form thorium 233; this isotope decays to protactinium 233, which in turn decays to the fissionable isotope uranium 233 which then fissions to produce heat energy.

Primary Air. Combustion air introduced into a furnace through the fuel bed by natural, induced or forced draught; or mixed with the fuel and delivered to the furnace under pressure. See **Secondary Air; Tertiary Air.**

Primary Burner. In respect of an incinerator a burner installed in the primary combustion chamber to dry out and ignite the material to be burned. See **Secondary Burner.**

Primary Heating. The heating of heavy viscous grades of oil in the storage tank in order to keep the oil at a viscosity at which it may be pumped; this can be done by steam or hot-water coils placed in the tank, or by electric heating elements. The minimum storage temperatures vary with viscosities. Examples of heating temperatures in relation to oil viscosities expressed in seconds Redwood No. 1 at 100° F (38° C) are: 150 to 400 seconds, 45° F (7·22° C); 400 to 2000 seconds, 65° F (18·3° C); 2000 to 4000 seconds, 75° F (23·9° C). See **Redwood Viscometer; Secondary Heating; Viscosity.**

Priming. (a) The entrainment of boiler water in the steam produced; priming may be due to excessively high water levels, or due to foaming as a result of excessive salts in the water; (b) In pumping, the replacement of residual air in a pump by the fluid being pumped.

Producer Gas. Gas manufactured by the action of air and steam on coke or coal. A *gas producer*, q.v., comprises a cylindrical water-cooled shell fitted with a fuel charging hopper at the top and a fire grate at the bottom equipped with air/steam blast tuyeres; this unit is frequently followed by gas cleaning plant. The composition of the gas produced depends on the proportion of water vapour in the blast and on the type of fuel gasified. A typical analysis for a coke produced gas, the constituents being expressed as percentages, is as follows: carbon monoxide, 29; hydrogen, 11; carbon dioxide, 5;

nitrogen, 55. The calorific value is in the range 130 to 140 Btu (72 to 78 Chu) ft³.

Projected Diameter. The diameter of a circle which has the same area as the projected profile of the particle.

Propane. C_3H_8. A paraffin hydrocarbon with a b.p. of $-44°$ F ($-41·4°$ C); sp. gr. of liquid at 60° F of 0·51; a gross calorific value of 2520 Btu/ft³ dry and 21,500 Btu/lb (net 2320 Btu/ft³ dry and 19,800 Btu/lb). The sulphur content is negligible with a maximum of 0·01 per cent by weight. There are 480 therms/ton (gross c.v.) and 24 ft³ air is required to burn 1 ft³ gas. Propane burns in air to carbon dioxide and water. Propane is available in cylinders, hence the term "bottled gas" suitable for domestic and commercial use. It may be used also for the enrichment of town gas; heating processes requiring special atmospheres; and the fuelling of fork-lift trucks. Under normal atmospheric conditions propane will gasify immediately while *butane*, q.v., requires pre-heating slightly. See **Liquefied Petroleum Gases.**

Propane De-Asphalter. An oil refining process which extracts feed for the fluid catalytic cracking unit from *residuum*, q.v., by its solubility in liquid propane.

Proportional Counter. A device for detecting charged particles and photons by the *ionization*, q.v., which they produce in a gas between two electrodes. See **Radiation Detector.**

Proportional Response Controller. An automatic device by which a proportion of the heat being supplied to a plant is cut out when the temperature exceeds a set point. This type of control ensures smaller temperature variations than with an *on-off control*, q.v. A technique known as "proportional response with reset" combines the two methods to give accurate control at the desired temperature.

Proration. A system in use in the United States and some other oil producing countries whereby production from every well (except new wells, or the largely exhausted "strippers") is limited to a given percentage of its capacity, and buyers are usually obliged to take the same percentage of production from all the wells connected to their gathering facilities. In the United States the amount of production allowed each month is usually prorated to market demand, as estimated by the competent State authority.

Proton. A fundamental particle which is a constituent of all atomic nuclei. It has a mass of $1·67248 \times 10^{-24}$g and a positive electric charge of $1·602 \times 10^{-19}$ *coulomb*, q.v. The number of protons in a *nucleus* represents the *atomic number* of the *element*, qq.v.

187

Proximate Analysis. The determination in a sample of fuel of: (a) moisture content; (b) volatile matter content; (c) ash content; (d) fixed carbon content. See **Adventitious Ash; Ash; Fixed Carbon; Free Moisture; Inherent Ash; Inherent Moisture; Ultimate Analysis; Volatile Matter.**

Psychometric Chart. A chart which defines the relationships between dry bulb temperature, wet bulb temperature, moisture content, relative humidity and total heat of air.

Pulverized Fuel. Fuel, usually coal, finely ground; the desirable degree of fineness is governed by the use to which the pulverized fuel is to be put. A standard for fuel to be fired in large watertube boilers is that 65 per cent of the pulverized fuel should pass through a 200 B.S. mesh with elimination of coarse particles above 30 mesh. The most suitable type of fuel for pulverizing is bituminous coal containing over 20 per cent of volatile matter and of not more than medium coking power; the ash content may be as high as 20 per cent. See **Pulverizer.**

Pulverized Fuel Ash (PFA). Finely divided greyish ash resulting from the combustion of pulverized coal. It has a large surface area of 2000 to 5000 cm^2/g and 75 per cent or more will pass through a 200 B.S. sieve. A typical percentage analysis for fly ash from Central Electricity Generating Board power stations is: silica, 42·6; alumina, 32·4; ferric oxide, 10·4; lime, 5·1; magnesia, 2·5; sulphate, 0·9; alkalis, etc. 1·1; loss on ignition, 5·0. The "loss on ignition" is due to unburnt carbon, a variable percentage depending on combustion conditions which may be as low as 1 per cent and may rise to 12 per cent; the average for British base load stations is about 4 per cent. Formerly regarded solely as a waste product, over one million tons a year of PFA is being used in Britain in building materials.

Pulverizer. A device for grinding fuel as finely as possible with a minimum expenditure of power. By reducing coal to a fine powder it can be carried into a furnace by an air blast. Pulverizers fall into three categories, slow, medium and high-speed types. See **Ball Mill; Ball-race Mill; High Speed Mill; Raymond Bowl Mill.**

Pumped Storage. A means of "storing" electricity in the form of water by pumping it to a high level when there is electricity available to drive pumps, perhaps during the night when electricity demand is low, and then letting the water flow down through water turbines and so generating electricity the following day when demand is high. It is a technique for helping to reduce the effects of fluctuating demand for electricity on an electricity supply system: the addition of a pumped storage station to a system saves the installation of an equivalent capacity of conventional generating plant that would otherwise be required to meet the peak demand. A pumped storage scheme is in operation at Ffestiniog, Wales, the upper reservoir being about 1000 ft above the lower reservoir. At peak load periods Ffestiniog station generates 300 MW.

Pure Coal. The "*dry, mineral-matter-free*", q.v., portion of coal; the basis to which all data and observations are referred when fundamental questions are being considered. The "*as received*", q.v., basis is adopted for practical and commercial considerations.

Pyrites. A hard, yellow sulphide of iron which occurs as an impurity in coal; "fool's gold". Pyritic sulphur may occur in lenses, bands, veins, joints, balls or fossils; or it may occur as finely disseminated particles. It may vary in size from particles a few microns in diameter up to lumps several feet in diameter.

Pyrolysis. The decomposition of a substance by heating.

Pyrometer. A device for measuring high temperatures. See **Light Sensitive Pyrometer; Optical Pyrometer; Radiation Pyrometer; Suction Pyrometer; Thermo-Electric Pyrometer; Total Radiation Pyrometer; Two-Colour Pyrometer; Venturi Pneumatic Pyrometer.** Also **Line Reversal Method; Thermometer.**

Pyrometric Cones. Small pyramids about 2 in. high made of selected mixtures of oxides and glass which soften and melt at known temperatures; they are widely used in the ceramic industry as a method of measuring high temperatures in refractory heating furnaces.

Q

Quarl. A refractory throat around a burner port; it helps direct air into the flame and radiation from the hot refractory also assists efficient combustion.

R

Rad. The unit of absorbed dose of radiation; one rad is equal to an energy absorption of 100 ergs per gramme of body tissue. One millirad $= 10^{-3}$ rad. See **Roentgen Equivalent, Man.**

Radial-tip Fan Blading. A design of blading usually preferred for centrifugal induced draught fans; it has a greater resistance to erosion and there is a lower tendency for grit in the gas stream to

Radial-tip Fan Blading

build up on the back of the impeller blades compared with *backward-curved fan blading*, q.v. See Fig. 21. See **Fan.**

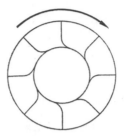

FIG. 21. *Radial-tip fan blading.*

Radiant Superheater. See **Superheater.**

Radiation. The transmission of heat from one point to another without affecting the temperature of the medium through which it passes. The heat dissipated by radiation from a surface may be determined by the formula:

$$Hr = 17.3 \times 10^{-10} E(T_1{}^4 - T_2{}^4)$$

where, Hr = radiation in Btu/ft^2/h;

T_1 = temperature of radiating surface in °R;
T_2 = temperature of surface receiving radiation in °R;
E = coefficient of emission.

The "coefficient of emission" is unity for a *black body* q.v.; other examples are: red brick, 0·94; refractory brickwork, 0·75 to 0·8; polished metals, 0·04 to 0·09. See **Heat Transfer.**

Radiation Absorber. A substance which absorbs atomic particles and radiation. The control of nuclear reactors is commonly effected by the use of control rods which contain materials, particularly boron and cadmium, which are good absorbers of thermal neutrons.

Radiation Detector. A device for detecting the presence of *radio-activity*, q.v. See **Geiger-Müller Counter; Ionization Chamber; Proportional Counter; Scintillation Counter.**

Radiation Pyrometer. A device for measuring temperatures, utilizing the emission of radiant energy from a hot body. It is suitable for measuring temperatures higher than those measured by a *thermoelectric pyrometer*, q.v., or for measuring the temperatures of

190

the inside of a furnace or kiln. There are three kinds of radiation pyrometer available: (a) *total radiation pyrometer*; (b) *optical pyrometer*; (c) *light sensitive pyrometer*, qq.v. See **Pyrometer; Thermometer.**

Radioactivity. The spontaneous emission of ionizing particles and rays following the disintegration of certain atomic nuclei. When a large number of radioactive atoms are present the random nature of individual emissions is obscured and the radiation appears to be regular and uniform. It can be defined as the number of disintegrations per unit time taking place in a radioactive specimen. A unit of radioactivity is the *curie*, q.v., which is 3.7×10^{10} (37 thousand million) disintegrations per second. The following sources of radioactive emissions to atmosphere are listed in *Meteorology and Atomic Energy*, United States Atomic Energy Commission, 1955.

(a) Mixing and handling uranium ores
(b) Chemical production of brown oxide or uranium (UO_2)
(c) Machining radioactive and toxic metals (e.g uranium and and beryllium)
(d) Atomic laboratories
(e) Particle accelerators
(f) Nuclear reactors
(g) Chemical processing plants for reactor fuels
(h) Waste disposal.

Plants handling radioactive materials are designed to prevent emissions of radioactive material to the atmosphere. Methods used for the removal of radioactive particles from gas streams are the same as those used for non-radioactive particulate matter except that the collection efficiency must be almost perfect.

Radiochemistry. The branch of chemistry concerned with the chemical processing of irradiated material or naturally radioactive material in order to isolate the various radioactive isotopes, and with the use of radioisotopes in the study of chemical problems.

Radiography. A method of detecting flaws, inclusions, lack of homogeneity, etc. in solid objects by making shadow images on photographic emulsions by the action of X-rays or gamma rays which have been differentially absorbed in their passage through the object.

Radiometer. An instrument for measuring the radiant heat emitted by an open fire. A test grate is raised above the floor so as to be at the centre of an imaginary hemisphere; the radiation falling on the inside surface of this hemisphere is measured by thermopiles. The readings are integrated to a value for total radiant energy.

Radiopotassium. The chief source of radioactivity inside the human body; about $\frac{1}{500}$ of body weight is due to potassium. About 1 potassium atom in every 8000 is K^{40} which is a radioactive isotope.

Radon. A radioactive gas which is released from the soil into the atmosphere. It is also released during the combustion of coal and for this reason there is more radon in the air over cities than in country air.

Ramsbottom Test. A test to determine the *carbon residue*, q.v., of a liquid fuel, which is generally preferred to the *Conradson Test*, q.v. A glass bulb containing a sample of oil is heated in a bath of molten metal maintained at 1022° F (550° C). After cooling the residue is weighed and expressed as a percentage of oil used.

Rank. In respect of coal, the degree of metamorphosis or "coalification" that the original plant debris has undergone during the geological ages since it was deposited. Lignites and brown coals are low-rank coals, while anthracite is of the highest rank.

Rankine Cycle. An ideal cycle for a steam engine, proposed by Professor W. J. M. Rankine (1820–1872) as a standard of comparison with the performance of actual engines. The cycle comprises four stages: (a) steam passes from the boiler to the cylinder at constant temperature; (b) steam expands adiabatically to the condenser pressure; (c) heat is given to condenser at constant temperature; (d) condensation is completed and condensate returned to boiler. In the Rankine Cycle the work done is equivalent to the total heat H_2 in the steam at the end of the adiabatic expansion, subtracted from the total heat H_1 in the steam at the beginning of the expansion. The heat supplied is equal to the sensible heat h_2 in the condensed steam subtracted from H_1. The cycle efficiency of the engine is calculated as follows:

$$\text{Efficiency} = \frac{\text{Work done}}{\text{Heat supplied}} = \frac{H_1 - H_2}{H_1 - h_2}$$

Rankine Scale. A temperature scale based on three fixed points: (a) the lowest (0°) being abolute zero; (b) the second (491·67°) being the temperature of melting ice; (c) the highest (671·67°) being the boiling point of water at normal atmospheric pressure. The Rankine and Kelvin scales are absolute scales, 0° R = 0° K; 1·8 Rankine degrees = 1 Kelvin degree. See **Absolute Zero; Temperature Scales.**

Rational Analysis. The resolution into chemical types of a mass of rock or coal. In respect of coal the constituents determined are:

(a) resins, waxes and extractable hydrocarbons; (b) cuticular plant remains; (c) humic substances; (d) opaque matter; (e) *fusain*, q.v.

Rational Classification. A system of coal classification based primarily on the *ultimate analysis* q.v. of the different chemical entities present in coal. See **Rational Analysis.**

Raw Gas. Gas before purification.

Raymond Bowl Mill. A medium-speed *pulverizer* for the production of *pulverized fuel,* qq.v. Raw coal enters a bowl rotating at between 74 to 150 rev/min; there it is ground between a ball ring and rollers. Hot air for drying and conveying the pulverized fuel enters the lower mill casing and a rapid circulation of hot air is maintained through the system. The powdered coal is drawn through a classifier where the over-sized particles are returned to the inlet of the mill. The outlet temperature aimed at is usually 160 to 180° F (71 to 82° C).

Reactive Fuel. See **Reactivity.**

Reactivity. In relation to a fuel, e.g. coke, the measure of ease of ignition from cold and of the relative rate of combustion under specified conditions. British Standard 3142 defines standards for domestic cokes.

Reactor. Combustion units in which a fuel and oxidizer are chemically combined to obtain a specific chemical compound, e.g., the combination of sulphur and oxygen to produce sulphur dioxide. In a "partial combustion" reactor a less than stoichiometric supply of air is used, as in the reaction of carbon with oxygen to produce carbon monoxide.

Reactor Kinetics. The branch of reactor technology concerned with the study of the behaviour of a reactor when its conditions of operation are not steady.

Réaumur Temperature Scale. A temperature scale in which the fixed points are 0° (freezing point of water) and 80° (boiling point of water); a little used scale.

Reciprocating Pump. A positive displacement pump consisting of a piston moving back and forth within a cylinder. With each stroke a definite volume of liquid is pushed out through the discharge valve.

Reclaimed Coal. *Coal,* q.v. recovered from a stockpile, as distinct from being received direct from a supplier.

Recorder. An instrument fitted with a chart, rotated by some form of clockwork or electric mechanism, upon which a pen draws a diagram indicating a measured change in an operating condition, e.g. temperature or smoke emission. The charts provide a permanent record.

Rectifier. A device for the conversion of an alternating current into a direct current by the inversion or suppression of alternate half-waves. See **Electronic Valve Rectifier; Electrostatic Precipitator; Mechanical Rectifier; Silicon Diode Rectifier; Static Rectifier.**

Recuperative Air Preheater. An *air preheater*, q.v., constructed of tubes or plates, the gases passing on one side and the air on the other, usually in a contra-flow manner in order to ensure a maximum heat transfer through the plates.

Red Top Burning. The condition of a fuel bed which is uniformly incandescent at its surface. See **Black Centre Burning.**

Reducing. In petroleum refining, the removal of light hydrocarbons by *fractional distillation*, q.v.

Reduction. The removal of oxygen from a compound; the reverse of *oxidation*, q.v. A reducing atmosphere is one devoid of free oxygen.

Redwood Viscometer. An instrument for measuring the *viscosity*, q.v., of fuel oils, expressed in Redwood seconds at 100° F (37·8° C). Ordinary fuels are measured in Redwood No. 1 seconds; more viscous oils in Redwood No. 2 seconds. The Redwood No. 2 cup has a larger orifice which allows the sample of oil to flow out in about one-tenth of the time taken in the Redwood No. 1 apparatus.

Refinery Capacity. In respect of oil refining, the maximum throughput of crude oil for which the plant is designed. A refinery's output of refined products will usually be about 5 per cent smaller than this, allowing for waste and for the products burned to keep the refinery and its auxiliary plant in operation.

Refinery Flares. A method of disposing of surplus gas at oil refineries by burning such releases at the top of a flare stack. Fuel gas cannot be released to the atmosphere unburnt partly because of odour and partly because of explosion hazards. The simple burning of refinery fuel at the top of a stack involves the production of considerable smoke; flares are made to a large extent smokeless either by steam injection or water spray close to the point of ignition. The adjustment of the steam or water to meet the requirements of the flare is usually by manual control.

Refinery Gas. Any form or mixture of gas gathered in a refinery from the various stills.

Reflector. A layer of material surrounding the core of *nuclear reactor*, q.v., whose purpose is to reduce the escape of neutrons by means of scattering processes which result in the return of many of the neutrons to the core; a reflector material should have a high neutron-scattering cross-section and a low neutron-capture cross-section.

Refluxing. In *fractional distillation*, q.v. in oil refining, the return of part of the distillate to the *distillation tower*, q.v., to assist in making a more complete separation into the required fractions; the material returned is the reflux.

Re-formed Gas. Low thermal value gas obtained by the *pyrolysis*, q.v., and steam decomposition of high thermal value gases, e.g. natural gas or oil-refinery gas.

Reformer. A catalytic refining unit used in oil refineries to upgrade low octane gasolines into high octane gasolines by rearrangement of the molecular structure. The catalyst usually contains platinum. Reformers generally yield by-product hydrogen.

Reformer Pretreater. A hydrogenation unit used in oil refineries for purifying the feed to reformers. The pretreater effectively removes catalyst poisons and protects the platinum catalyst in the *reformer*, q.v.

Refractory. A material used in lining furnaces; it must be capable of resisting a moderate load at high temperatures without distortion or collapse, changes in temperature and the action of molten metals, slags and hot dusty gases. See **Alumina; Aluminous Firebrick; Chrome Brick; Chrome-magnesite Brick; Dolomite Brick; Firebrick; Hot Face Insulation; Magnesite Brick; Semi-siliceous Brick; Silica; Silica Brick; Siliceous Brick; Sillimanite Brick.**

Refrigeration. The cooling of air or liquid by passing it over coils containing a cooling medium which can be refrigerant, chilled water or brine. Chilled water and brine are generated from a liquid chilling machine which has a refrigerant cycle; the water or brine is pumped in a closed circuit to the air conditioning or process equipment, and then back to the chiller for re-cooling. Ammonia, carbon dioxide, sulphur dioxide, methyl chloride and *freons*, q.v., are all possible refrigerants, although the freons have proved most popular because of their special characteristics. See **Refrigeration Plant.**

Refrigeration Plant. Plant specifically designed to achieve *refrigeration*, q.v., and comprising four basic components: (a) an evaporator, where heat is absorbed from the medium to be cooled (either air or liquid) by the evaporation of the liquid refrigerant; (b) a compressor, which raises the temperature and pressure of the refrigerant gas from the evaporator by compressing it; (c) a condenser, which is a heat transfer coil, where the hot gas is cooled by the condensing medium; (d) an expansion valve, which reduces the pressure of the liquid refrigerant to a point where it will vaporize at low temperature in the evaporator. Figure 22 gives a diagrammatic presentation of the refrigeration cycle.

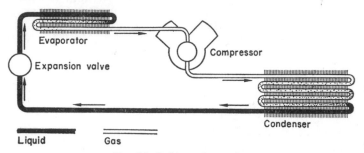

Fig. 22. *Refrigeration cycle.*

Regeneration. In catalytic processes in oil refineries, the burning of the deposits on the catalyst with an oxygen containing gas.

Regenerative Condenser. A steam condenser which in addition to condensing the steam, heats the condensate.

Regenerative Feed-heating. The use of steam bled from a *steam turbine*, q.v., to heat the feed water intended for the boiler.

Regenerator. A *heat exchanger* consisting of a mass of *chequer-brickwork*, qq.v., or other heat absorbing solid constructed or placed in a chamber; during the heating stage, hot waste gases from the furnace pass through the chamber and heat the chequers, escaping via a reversing valve to the chimney. After a prescribed interval the direction of flow in the chamber is reversed and cold combustion air drawn in, acquiring the stored heat of the chequers and passing in a preheated condition to the furnace. In glass-melting and open-hearth steel furnaces, for example, regenerators of this type are employed, a reversal of gases taking place once every 20 minutes or so.

Reheat. The passing of the exhaust steam from the high pressure cylinder of a turbine through a reheater section of the boiler; there it is reheated usually to its original temperature before passing to the intermediate-pressure cylinder of the turbine.

Reid Vapour Pressure (R.V.P.). An important specification for gasoline; it is a measure of the vapour pressure of a sample at 100° F (38° C). The test is usually made in a "bomb" and the results are reported in pounds per square inch. Low-boiling hydrocarbons have a high vapour pressure.

Reinluft Process. A process for removing sulphur dioxide from flue gases, involving the use of a falling bed of activated carbon through which the flue gases are passed, SO_2 and SO_3 being adsorbed. The sulphur dioxide gas is subsequently recovered by desorption at a higher temperature, when the adsorbent is regenerated without loss of activity. See **Battersea Gas Washing Process; Howden-I.C.I. Process; Fulham-Simon-Carves Process.**

Relative Biological Effectiveness (R.B.E.). The inverse ratio of tissue doses of two different kinds of radiation which produce the same biological effect.

Relative Humidity. The ratio of the amount of water vapour in the atmosphere to the amount which would saturate it at the same temperature; or the ratio of the actual vapour pressure of the water vapour to the maximum or saturated vapour pressure of the water vapour at the same temperature. At *dew point*, q.v., the relative humidity is 100 per cent; a rise of temperature without the addition of more vapour lowers the relative humidity, the *absolute humidity*, q.v., remaining the same, while a fall in temperature raises it. A hygrometer is used to measure relative humidity.

Rerun Oil. In a petroleum refinery, oil which has been redistilled.

Research Method. A test to determine the *octane number* of a *gasoline*, qq.v., utilizing an engine speed of 600 rev/min and a mixture temperature of 120° F (49° C). The result gives good agreement with engines operating under mild conditions with a low engine speed; it is a less severe test than that offered by the *motor method*, q.v.

Residence-time. The time taken by an element of fuel to pass through a combustion chamber; the time depends on the particular path that the element has followed in its passage through the chamber, the possible paths depending on the flow pattern within the chamber.

Residual. See **Residuum.**

Residual Fuel. Any liquid fuel containing the *residuum*, q.v., from crude distillation or thermal cracking. A typical *Central Electricity Generating Board*, q.v., specification for a residual fuel oil is: specific gravity at 60° F (15° C), 0·98; viscosity Redwood No. 1 at 100° F (38° C), 5930 seconds; closed flash point, 200° F (93° C); gross calorific value, 18,300 Btu/lb (10,165 Chu/lb); sulphur, 4·2 per cent; pour point, 30° F (−1° C). See **Viscosity.**

Residuum. The most non-volatile portion of petroleum; residuums

197

are sometimes described as long or short residuums. A long residuum is the non-volatile residue from an atmospheric pressure distillation; a short residuum is obtained from vacuum distillation.

Residuum Hydro Desulphuriser. An oil refinery unit utilizing a special catalyst and a fairly high hydrogen partial pressure. Hydrogen is consumed in the process and sulphur is released as H_2S.

Resistance. The resistance which an electrical conductor or appliance offers to the passage of an electric current. It is measured in ohms. See **Ohm.**

Resistance Heating. The production of heat by direct application of a voltage to a resistor; the efficiency of conversion of electrical energy into heat by this method is 100 per cent. Resistor material may be an alloy of nickel and chromium, graphite, molybdenum or silicon carbide.

Resistance Thermometer. A device for measuring temperature. It depends upon the known variation of electrical resistance of a metal wire as the temperature changes. The detection element is a wire of platinum or nickel; the measuring element is an electrical instrument for measuring the electrical resistance of the detecting element and registering this in terms of temperature. The resistance thermometer is capable of measuring temperatures from $-400°$ F to $+1832°$ F (-240 to $+1000°$ C) but, for industrial use, it is not recommended for temperatures above $1112°$ F ($600°$ C). See **Pyrometer; Thermometer.**

Restricted Hour Tariff. A *tariff*, q.v., offering low rates for supplies which are restricted automatically by time switches to certain off-peak hours of the day.

Reversion Pressure Test Burner. A test burner which assesses the tendency of a non-aerated flame to " lift " or move away from the burner head. The gas supply pressure to the burner is increased until the flame lifts and is then gradually reduced until the flame returns to the burner head; this pressure is the reversion pressure. A high reversion pressure ensures flame stability; town gas often has a reversion pressure of between 4 and 5 in w.g.

"Rexco ". A reactive coke produced under "low temperature" carbonization conditions when coal is heated in retorts to a temperature of the order of $1120°$ F ($600°$ C); it is produced by the National Carbonizing Co. Ltd. It ignites readily, burns with very little smoke emission, and is well-suited for all domestic appliances including old-fashioned stool-bottom grates. The volatile content is about 7 per cent with a bulk density of about 30 lb/ft^3. It is an

authorized fuel in a *smoke control area*, q.v., See **Authorized Fuels;**
"Coalite"; **Coke.**

Reynolds Number. A non-dimensional ratio which defines the type
of flow occurring in a pipe. It is calculated:

$$R_e = \frac{vdp}{n}$$

where, R_e = Reynolds Number
v = velocity of fluid
d = pipe diameter
p = density of fluid
n = absolute viscosity.

Any self-consistent units may be used. Generally, streamline
flow = $R_e < 2000$; transitional flow = R_e 2000 to 4000; turbulent
flow = $R_e > 4000$.

Rich Gas. Gas of high calorific value, exceeding 900 Btu/ft³
(500 Chu/ft³). See **Lean Gas.**

Rich Oil. *Absorption oil*, q.v., containing dissolved natural
gasoline fractions.

Riddlings. Unburnt pieces of solid fuel which fall between the
firebars of a furnace.

Ridley Report. The Report of the Committee on National Policy
for the use of Fuel and Power Resources (Cmnd. 8647, 1952).

Ring Circuit. An electrical circuit or main arranged in the form of
a ring or loop as distinct from a single, open-ended length; it is used
in domestic situations to provide an adequate number of socket
outlets of a 13 amp universal type at a lower cost than the traditional
system of radial wiring to each point.

Ringelmann Chart. A shade chart, devised over sixty years ago by
Professor Ringelmann of France, used for the observation from a
distance of the density of smoke issuing from a chimney. The
estimation is made by comparing the shade of the smoke against
shade cards held some 50 feet from the observer and in line with the
chimney; the shade cards consist of grills of black lines, each grill
being 4 inches square. The shades are:

Number	Description	Approx. % black on shade card
1	Light grey	20
2	Darker grey	40
3	Very dark grey	60
4	Black	80

Ringelmann Chart

Clean air legislation frequently prohibits the emission of "dark smoke" from chimneys, save for short periods permitted by Regulation. Dark smoke is usually defined as "smoke which, if compared in the appropriate manner with a chart of the type known as the Ringelmann Chart, appears to be as dark as or darker than shade 2

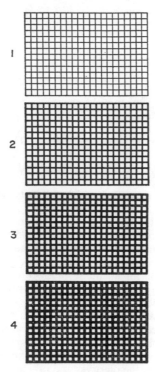

Fig. 23. *Ringelmann chart.*

on the chart." British Standard 2742 : 1958 describes the use of the Chart in detail; B.S. 2742M describes the use of a miniature smoke chart. Figure 23 shows a reduced reproduction of a Ringelmann Chart.

Ring Main. (a) A system for supplying *pulverized fuel*, q.v., to boilers or furnaces. It consists of a pulverizer of the slow, medium or high-speed type, from which pulverized coal is extracted and delivered by an exhauster fan to a cyclone situated above a bin; this cyclone extracts the coal dust which is deposited in the bin. Feeders at the base of the bin supply coal to ring mains which encircle the shop in which the furnaces or boilers to be fired are situated. Pulverized fuel not consumed is returned to the cyclone for redistribution. Ring mains vary in length up to 1000 ft.; they enable batteries of medium sized furnaces to be serviced. (b) A system for supplying oil to a number of burners fitted to boilers or furnaces; oil is pumped from a storage tank, preheated if necessary, and passed through an oil heater in which it is raised to the correct atomizing temperature. From the heater the oil is circulated through a ring main to which branch pipes serving the individual burners are connected. Through continuous circulation the oil is kept at the correct temperature.

Ritchie Boiler. See **Thermal Storage Boiler.**

Rittinger's Law. A statement that the work required to produce material of a given size from a larger size is proportional to the area of the new surface produced.

Roentgen, (r). A unit of radiological dose. It is the quantity of X-or gamma radiation capable of liberating ions carrying 1 electrostatic unit of charge of each sign in 0·001293 grammes of air (equivalent to 1 cm³ of dry air at 32° F (0° C) and 760 mmHg). This is equivalent to the release of 83·8 ergs of energy in 1 gramme of air. The megaroentgen (Mr) = 10^6 roentgens; the kiloroentgen (kr) = 10^3 roentgens; and the milliroentgen (mr) = 10^{-3} roentgen.

Roentgen Equivalent, Man. (rem). The quantity of any ionizing radiation such that the energy imparted to 1 gramme of tissue has the same biological effect as an absorbed dose of 1 *rad*, q.v., of X-radiation. It follows that:

$$\text{Dose in rems} = (\text{dose in rads}) \times (\text{relative biological effectiveness}).$$

Roentgen Equivalent, Physical. (rep). A unit of absorbed dose which has now been replaced by the *rad*, q.v.

Roga Index. An index of the caking properties of a *coal*, q.v., it is determined in a laboratory tumbler test of a coke button which has been made by heating a mixture of the coal under consideration, and anthracite.

Roller Filter. An *air filter*, q.v., consisting of a roll of suitable fabric. As particulate matter is collected on the exposed area, the roll is wound on bringing clean areas of the filter medium into use. Collecting efficiencies of up to 99 per cent are claimed in the 1 to 3 micron particle size range.

"Roomheat". A cushion-shaped fluidized char binderless briquette which burns with a long flame giving an attractive fire; it is suitable for room heaters and open domestic fires. It is manufactured by the *National Coal Board*, q.v., at Markham, near Doncaster, England. See **Authorized Fuels; Smoke Control Area.**

Rosin-Rammler Equation. A formula for expressing the size distribution of powders over a range from 2 to 98 per cent of the whole; it is based on the fact that the distribution of particle sizes in a natural or artificial powder tends to follow an exponential law:

$$R = 100e^{-bx^n}$$

where, R = percentage of powder resting on a sieve having an aperture of x inches

b and n = constants for any given powder.

For *pulverized fuel*, q.v., produced under normal conditions b ranges from 120 to 800, and n ranges from 0·8 to 1·1. The formula is discussed in a paper by Rosin and Rammler, *J. Inst. F.*, 1933–34, **7**, 29. The formula may be applied graphically by using special logarithmic paper, as discussed by Bennett, *J. Inst. F.*, 1933, 34, **7**, 109.

Rotary Burner. Or spinning cup burner. A low pressure air burner in which oil is distributed over the air blast from the lip of a rapidly rotating cup; the cup is rotated by an air turbine or an electric motor See **Oil Burner.**

Rotometer. An instrument for measuring the rate of fluid flow. It consists of a tapered vertical tube with circular cross-section containing a float which is free to move in a vertical path to a height dependent upon the rate of fluid flow upward through the tube. It is based on the principle of *Stokes' Law*, q.v.

Ruhrgas Gasifier. A *cyclone gasifier*, q.v., for gasifying coal fines to make producer gas. Developed by Ruhrgas A. G.

Run-of-mine Coal. Coal as raised from the mine or pit, before screening.

Run-of-Retort Coke. Or run-of-oven coke. Coke from a retort or oven before undergoing screening or other process.

Rutherford. (rd). A unit of radioactivity. It is the quantity of any radioactive nuclide in which the number of disintegrations per second is 10^6.

Safety Rod. A neutron-absorbing rod which can be inserted rapidly into the core of a *nuclear reactor*, q.v., in the event of an emergency, thus enabling the reactor to be shut down.

Safety Valve. A device to prevent a boiler from working in excess of the "maximum permissible working pressure" by automatically discharging to the atmosphere the excess steam generated in the boiler. Each valve should be large enough for its purpose and all industrial boilers, with the exception perhaps of the small vertical boiler, should be fitted with at least two safety valves. Thus if one valve fails the other will function. Each safety valve should be tested daily to ensure that it will blow at the correct pressure. See **Dead-weight Safety Valve; Lever or Steelyard Safety Valve; Spring-loaded Safety Valve.**

Samarium, Sm. An *element*, q.v.; at. no. 62; at. wt. 150·35. Samarium is produced in a *nuclear reactor*, q.v.; samarium 149 has a high neutron-capture cross-section. See **Fission Products.**

Sampling Probe. A tube inserted in a chimney or duct in order to draw off into measuring equipment a sample of gas.

Sankey Diagram. A heat flow diagram for an industrial process in which the quantity of heat in the various items of the *heat balance*, q.v., for the plant is represented by the width of a band. Figure 24 shows a Sankey Diagram for an oil-fired *economic boiler*, q.v., indicating the ingoing and outgoing flows of heat; the widths of the bands are proportional to the amounts of heat represented.

Sapropel. Initially a slimy, putrefying, sediment formed in deep, stagnant waters where the remains of minute organisms, algae, spores, pollen and plant fragments accumulate in virtually complete absence of air; on consolidation sapropel contains a higher proportion of hydrocarbon-rich compounds. The sapropelic coals used to be a valuable source of gas. See **Boghead Coal; Cannel Coal.**

Saturated Steam. Steam which has taken up its full quota of latent heat, containing no moisture or suspended unevaporated water; synonymous with *dry saturated steam*, q.v.

Savannah, N.S. The world's first nuclear-powered merchant ship, launched on 21st July, 1959; a United States test ship built to prove the feasibility of nuclear power for merchant ships. The reactor system is a pressurized water type consisting of a single reactor with

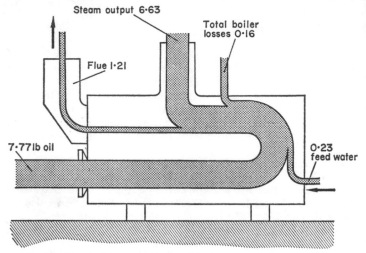

Fig. 24. *Simple diagrammatic heat-flow diagram or Sankey diagram.*

two main coolant loops and two steam boilers. The reactor pressure is 1735 lb/in² with a reactor water flow of 8×10^6 lb/h. The uranium oxide fuel in the form of pressed sintered pellets is contained in stainless steel tubes; each of the 32 fuel elements contains 164 of these tubes. The reactor is controlled by stainless steel rods containing enriched boron.

Saybolt Furol Viscometer. A U.S. instrument for measuring the viscosities of more viscous oils; it is based on the time in seconds required for a unit volume of oil to pass through an aperture at a prescribed temperature.

Saybolt Universal Viscometer. A U.S. instrument for measuring the viscosities of oils; it is based on the time in seconds required for a unit volume of oil to pass through an aperture at a prescribed temperature.

S.C.A. Steel-cored-aluminium conductor used in the transmission system operated by the *Central Electricity Generating Board*, q.v.

Scaling. A condition in which the thickness of oxides and corrosion products reaches a certain value, the layer cracking and falling away from the metal often leaving it exposed to fresh attack.

Scarfing. The use of oxygen flame jets to remove surface defects from steel.

Scintillation Counter. A device for detecting charged particles or photons by the flashes of light (scintillations) which they produce in certain materials known as phosphors, e.g., sodium iodide and anthracene. The light falls on the cathode of a photomultiplier which converts each scintillation into a pulse of current and amplifies the latter so that it can operate associated counting equipment. See **Radiation Detector.**

Scotch Marine Boiler. See **Marine Boiler.**

Scraper Conveyor. A conveyor which removes ash from the back of a furnace, bringing it to the front where it is discharged into an ash shute.

Screening. The passing of coal over bars, perforated plates, or wire mesh screens, so that sizes smaller than the openings fall through. See **Cylindrical Screen; Shaking Screen; Trommel; Vibrating Screen.**

Seasonal Tariff. A *tariff*, q.v., for the supply of electricity in which a higher price per unit of electricity applies during the winter months than during the summer months.

"Sebrite". A coke meeting British Standard 3142; it is produced by the South Eastern Gas Board of England in intermittent vertical retorts. Peat is used in its manufacture which has the effect of producing a reactive coke. It is suitable for domestic fires, including approved inset grates. See **Intermittent Vertical Retort.**

Secondary Air. Combustion air introduced above or beyond a fuel bed by natural, induced or forced draught; it includes all "overfire" air which may be introduced through the front or side walls, or through the bridge wall, of a furnace. Fuels containing a very high proportion of *volatile matter*, q.v., require the greater part of the total weight of air required for combustion to be supplied over the grate. The burning of *wood*, q.v., is an example of this. On the other hand, *coke*, q.v., with a very low volatile content requires only enough secondary air to burn the *carbon monoxide* to *carbon dioxide*, qq.v. See **Overfire Jets; Primary Air; Tertiary Air.**

Secondary Burner. In respect of an incinerator a burner installed in the secondary combustion chamber to maintain a minimum temperature of about 1400° F (760° C). See **Primary Burner.**

Secondary Heating. The heating of oil to bring it to the correct temperature for efficient *atomization*, q.v. This is achieved by heating the oil with steam or electricity, with thermostatic control. With all plants and grades of oil the temperature of the oil is raised to that at which its viscosity is between 80 and 100 seconds Redwood No. 1 at the burner. Examples of heating temperatures in relation to the desired oil viscosities are: (a) light fuel oil, 120 to 140° F (49 to 60° C); (b) medium fuel oil, 175 to 200° F (79 to 93° C); (c) heavy fuel oil, 250° F (121° C). See **Primary Heating; Viscosity.**

Sectional Boiler. Cast-iron or steel hot-water boiler, made up in sections which enable it to be increased or decreased in size; most widely used for central heating. Cast-iron boilers are most common and can be used in buildings where the working head does not exceed 120 ft; above 120 ft steel boilers are necessary. Sectional boilers can be supplied with capacities up to about 3,500,000 Btu/h (1,900,000 Chu/h) to supply a building of up to 1,000,000 ft^3 from a single boiler. Sectional boilers have a life of about 20 years and can be operated with efficiencies of up to 70 per cent with mechanical firing. See **Boiler.**

Sedimentation. The determination of the terminal velocities of particles by introducing a known weight of dust into a liquid, generally water or alcohol, and measuring the amounts of dust falling out at predetermined times.

Seeboard Process. A wet scrubbing process for the removal of *hydrogen sulphide*, q.v., from refinery and petroleum oil gas streams. The gas is scrubbed with a solution of sodium carbonate, Na_2CO_3. The dissolved hydrogen sulphide is subsequently removed by blowing air through the solution. See **Hydrogen Sulphide Removal.**

Segas Process. A cyclic process developed by Britain's South Eastern Gas Board for the catalytic manufacture of gas from oil; a lime catalyst is used to facilitate the reactions. Feedstock varying from liquid butane to residual oils can be used. In the blow stage, counterflowing air is blown through the air preheater, burning off any carbon; the preheated air brings the catalyst to a working temperature of 1450 to 1800° F (788 to 982° C) and also heats up the vaporizer and steam preheater, leaving the system via a waste heat boiler. A steam purge follows, in the same direction. In the make stage steam is admitted in the opposite direction to the blow. It passes through the preheater and then meets a counter-current oil spray; the steam/vaporized oil mixture then passes through the catalyst where the desired reactions take place. The constituents of a typical gas produced, expressed in percentages, are: Hydrogen, 50; Methane, 16; Carbon monoxide, 15; Carbon dioxide, 9; Hydrocarbons, 7; Nitrogen, 3. The calorific value is about 500 Btu/ft^3 (278 Chu/ft^3).

Segregation. The tendency of coal when poured for the larger pieces to separate themselves; thus when non-graded coal is poured on to a cone-shaped heap, the larger pieces run down the sides of the cone. Segregation may occur when there is transfer of coal from one conveyor to another and in bunkers. It can be reduced by fitting a flat plate above the outlet from bunkers, and by using travelling chutes to feed coal into hoppers.

Seismic Reflection Method. A method used in the exploration for oil. A charge of dynamite is exploded in the ground, usually within about 100 ft of the surface; the shock waves travel through the ground and bounce back from underground layers of rock. Seismometers placed at intervals on the surface receive the shock waves and transmit them to a recording instrument, where they are recorded on a seismogram. From these records, geophysicists determine the position and shape of anticlines and other formations lying deep in the earth. See **Petroleum.**

Selenium. Se. An *element*, q.v., used in the making of photoelectric cells; at. no. 34; at. wt. 78·96. See **Photoelectric Cell.**

Self-proportioning Burner. A low pressure air blast burner, supplying all the air required for combustion through the burner, and maintaining a correct air/fuel ratio over its complete range of fuel output; the oil and air flow rates are adjusted simultaneously by the movement of a single lever. See **Oil Burner.**

Semet-Solvay Process. A cyclic oil gasification process using thermal cracking in steam. The plant consists of two carburettors or gas generators connected at the top. During the "blow" period *primary air*, q.v., passes through one generator burning off carbon previously deposited on the *chequer-brickwork*, q.v., and enters the second chamber in which oil is burned in the presence of *secondary air*, q.v. During the "make" period cracking oil is injected into both generators from the top with steam passing through both chambers in the same direction as in the "blow" period. The process is then repeated with reverse directions for air and steam. Heavy oil is used in the process with a cracking temperature of 1200 to 1650° F (650 to 900° C). The *calorific value*, q.v., of the gas produced is from 1000 to 1200 Btu/ft³; typical composition is as follows, the constituents being expressed as percentages by volume: C_nH_{2n+2}, 40; unsats., 30; H_2, 18; N_2, 6·5; CO, 3; CO_2, 2·5.

Semi-siliceous Brick. A *refractory*, q.v., containing 78 to 85 per cent of silica, the balance being mainly alumina.

Sensible Heat. Heat, the intensity of which can be measured by thermometric techniques; as distinct from latent (or hidden) heat which produces no change in temperature. The heat required to raise the temperature of water from freezing point to boiling-point is sensible heat. See **Latent Heat; Total Heat.**

Sensitive Flame. A gas *flame*, q.v., which changes in shape when sound waves fall upon it.

Servomechanism. A device for amplifying by electronic or other means a small impulse from a measuring instrument; the larger force is then capable of operating a valve or other mechanism.

Settlement Chamber. A dust-arresting device for use in conjunction with boilers and furnaces; it consists of a rectangular chamber or an enlargement of a flue or duct, the effect being to reduce the velocity of the gases allowing grit to settle out. The chamber may contain a system of baffles to deflect the gases and assist in the removal of grit. Devices of this nature are only effective in respect of particles larger than, say, 100μ.

Severity. A term which describes the manner in which an engine rates a fuel on the road. A "severe" engine will rate a fuel near to its motor method octane number, while a "mild" engine will give a road octane number near to its research method octane number; modern overhead-valve engines are in the latter category, while side-valve engines tend to be severe. See **Octane Number.**

Sewage Gas. *Methane*, q.v., produced naturally during the process of digestion in sludge tanks at sewage works. In large works it is utilized for heating the tanks to accelerate the sludge digestion process, power generation, space heating and lighting, hot water supplies and laboratory work. See **Natural Gas.**

Seyler's Classification. A classification system for coals; the positions of all coals are plotted in a diagram using the percentages of (a) total carbon, (b) hydrogen, as the main parameters and co-ordinates. As all coals with a specified *volatile matter*, q.v., content occupy positions along a straight line this led to the construction of a series of nearly parallel "isovols". Similarly, all coals having any specified *calorific value*, q.v., occupy positions along a straight line and thus a series of roughly parallel "isocals" were constructed. The isovols and isocals are approximately at right angles to each other and inclined to the carbon and hydrogen coordinates. See **Coal Classification Systems.**

Shaft Mine. A mine in which the seams are reached by a vertical shaft from the surface.

Shaking Screen. A horizontal rectangular screen given a reciprocating motion in a lengthwise direction by means of an eccentric crank operating at 80 to 120 rev/min.

Shale Oil. Produced by the *destructive distillation*, q.v., of oil shale; crushed shale is placed in retorts and the organic material of shale, known as kerogen, is cracked with gas or steam at 662 to 932° F (350 to 500° C). The crude oil produced is similar to crude petroleum but contains higher percentages of sulphur, nitrogen and oxygen; the refining of shale oil is similar to that of crude petroleum. The average yield of a ton of shale is about 25 gallons of crude oil although this may vary considerably; 90 gallons of ammonia liquor, and a variable quantity of gas used to heat the retorts.

Shatter Test. Test for the impact hardness of coke; 50 lb of coke over 2 in. size is dropped four times from a height of six feet on to a metal plate. A sieve analysis follows. A good metallurgical coke should show 75 per cent over 2 in. in size; 85 per cent over 1½ in.

Shell Gasification Process. A process consisting of the non-catalytic, partial oxidation of hydrocarbon feedstocks using oxygen and steam under pressures of 150 to 600 lb/in² and at temperatures of about 2500° F (1370° C). A water gas is produced which can be used for a variety of purposes.

Shell Phosphate Process. A wet scrubbing process for the removal of *hydrogen sulphide*, q.v., from refinery and petroleum oil gas streams. The scrubbing medium is a solution of tri-potassium phosphate which removes hydrogen sulphide in the following reaction:

$$K_3PO_4 + H_2S \rightarrow K_2HPO_4 + KHS$$

The hydrogen sulphide is regenerated by boiling; the sulphur may be recovered in a *Claus kiln*, q.v. See **Hydrogen Sulphide Removal.**

Shell Smoke Meter. An instrument for measuring stack solids developed by the Shell Oil Company for use with oil-fired installations. It comprises a small motor-driven vacuum pump which draws a sample of the flue gases through a filter paper for a period of one minute, the pressure drop across the paper being maintained constant at 3 inHg; the timing is automatic. The stain on the filter paper is then compared with a scale of shades graduated from 0 (white) to 9 (black); the nearest matching shade is known as the Shell Smoke Number. Shade 9 is the equivalent to chimney emissions of less than shade 1 on the *Ringelmann Chart*, q.v., while shade 6 represents the lowest concentration of stack solids which is just visible to the human eye. The results obtained by this instrument are influenced by the type of fuel and size of smoke particle; a given weight of very fine particles has a much greater blackening effect than a similar weight of relatively coarse particles.

Short-circuit. An electric current taking a shorter path than intended; as the resistance of the unintended path is usually low, a rush of current takes place and blows the *fuse*, q.v.

Short Flame Coal. *Coal*, of low *volatile matter*, qq.v. See **Long Flame Coal.**

Shot-rain Cleaning. A method of cleaning fouled furnace tube surfaces by discharging or "raining" a large quantity of steel shot over the surfaces periodically.

Shuttle Kiln. A thin-walled, steel-encased rectangular intermittent kiln in use in the United States ceramic industry. There are two types: (a) the envelope kiln which has two fixed bases on which the ware is set and a movable cover on wheels which is pushed on rails over the base being fired; (b) the car shuttle kiln, which has fixed walls and crown of thin wall construction and cars set with ware are pushed into the kiln for firing.

Siegert Formula. A formula for calculating the heat loss in flue gases passing through a chimney:

$$S = \frac{{}^{1}K\,(t_2 - t_1)}{L}$$

where S = percentage of the *net calorific value*, q.v., lost in the flue gases;

t_1 = ambient air temperature, °F;

t_2 = flue gas temperature, °F;

L = percentage CO_2 content of the flue gases;

K = a constant equal to: anthracite 0·37; bituminous 0·35; coke 0·39; oil 0·31.

Sieving. A relatively simple and quick way of obtaining a sizing or grading analysis of a sample of dust or solid fuel; it is only practicable down to a particle size of about 50 microns. Table 8 gives details of B.S.S. Normal Test Sieves.

Sigma Calorimeter. A continuous recording *gas calorimeter*, q.v., based upon the differential expansion of two concentric steel tubes. One of the tubes is heated by the combustion of the gas under test; the other is cooled by the flow of combustion air. The differential expansion of the tubes is transmitted to a pen recorder, calibrated in Btu/ft³.

TABLE 8—B.S.S. NORMAL TEST SIEVES

Nominal mesh number	Aperture microns	Nominal mesh number	Aperture microns
8	2057	52	295
10	1676	60	251
12	1405	72	211
14	1204	85	178
16	1003	100	152
18	853	120	124
22	699	150	104
25	599	170	89
30	500	200	76
36	422	240	66
44	353	300	53

Silica, SiO$_2$. The dioxide of silicon; it occurs in crystalline form as quartz, cristobalite and tridymite and is an essential constituent of the silicate groups of minerals. Silica and silicates constitute an important part of the mineral impurities in coal. See **Acid Refractories.**

Silica Brick. A *refractory*, q.v., containing at least 92 per cent of silica.

Siliceous Brick. A *refractory*, q.v., containing from 85 to 92 per cent of silica, the balance being mainly alumina.

Silicon Diode Rectifier. A transformer-rectifier for converting the alternating current of normal electricity supply into high tension direct current electricity; this design has found use in electrostatic precipitators. It is claimed that these rectifiers operate at 98 per cent efficiency. They have no moving parts. See **Electrostatic Precipitator; Rectifier.**

Sillimanite Brick. A *refractory*, q.v., containing 55 to 65 per cent of alumina, and 25 to 35 per cent of silica.

Single-phase Power. See **Three-phase Power.**

Single Pole. A description applied when a switch or fuse is inserted in only one of a pair of wires in an electrical circuit; the switch or fuse is described as a " single pole " type. See **Double Pole.**

Sintering. The fritting together of small particles to form larger particles; the conversion of fine dust into hard agglomerates. Sintering

Sintering

is employed in the steel industry to effect a chemical and physical improvement of ores before being charged to the blast furnaces; it is a process in which fines are mixed with coke breeze and passed through a sintering furnace. A sintering furnace consists of an endless strand of travelling perforated pallets on which the mixture is spread; this moving grate may be some 6 to 10 ft in width and up to 120 ft in length. The mixture ignites as it passes under a firebrick arch, jets of burning gas being directed upon it. Air passes downward through the strand bed and the sinter mix is completely burned through by the time it reaches the end of the strand; upon cooling, the sinter is broken and graded. The gases from the strand, discharge end, breakers and screen require de-dusting before discharge to atmosphere. Large quantities of sulphur dioxide are emitted and chimneys of up to 350 ft or more are often essential.

Sized Coals. See **Graded Coals.**

Size Distribution. See **Size Fraction.**

Size Fraction. A portion of a powder, dust or fuel sample composed of particles or lumps between two given size limits. The distribution of the size fractions in the total sample is known as the "size distribution".

Size Stability. The ability of a coal to withstand breakage during handling and shipping. In one test, size stability is determined by twice dropping a 50 lb sample of coal from a height of 6 ft on to a steel plate. From the size distribution before and after the test, the size stability is reported as a percentage factor.

Skimming Plant. An oil refinery designed to remove and finish only the lighter constituents from the crude oil, such as gasoline and kerosine; the balance is usually sold as fuel oil.

Slab. Semi-finished steel rectangular in cross-section, with the width greater than twice its thickness.

Slag. Melted ash which sticks to the furnace walls and other parts of furnaces.

Slag-tap Furnace. A furnace in which the ash is deposited and leaves the furnace in a molten condition. From 30 to 50 per cent of the ash leaves through the ash hopper. Also known as a wet-bottom furnace. See **Dry-bottom Furnace.**

Slagging Gasifier. A gasifier in which the ash melts and is withdrawn in liquid form.

Sliding Damper. An adjustable plate normally installed horizontally or vertically in the flue between furnace and stack. See **Damper.**

Slip Coupling. A type of coupling between a motor and a fan; the speed of the fan is varied by altering the slip between the two coupling halves either magnetically or hydraulically.

Slope Mine. A mine similar to a *drift mine*, q.v., but in which the seams are at a perceptible angle to the horizontal line of entry.

Slug. An engineering unit of mass. One slug = 32·174 pounds.

Slurry Synthesis Process. A variant of the *Fischer-Tropsch synthesis process*, q.v., in which the catalyst is maintained in suspension in an oil.

Smalls. All coal down to dust that is smaller than a certain size. For example, one inch smalls consist of all the coal which passes through a screen with one inch holes.

Smithells Separator. A classical device for demonstrating the principle and nature of two-stage combustion. The jet of fuel gas entrains a fraction of its combustion air and burns sub-stoichiometrically at the top of the Bunsen tube; the hot products contain carbon monoxide, hydrogen and other intermediate gases which burn as a diffusion flame at the top of the quartz tube in the ambient air. See Fig. 25. See **Stoichiometric**.

Smog. Originally a term applied solely to a mixture of smoke and fog; it was first used by the late Dr. H. A. Des Voeux, founder-president of the National Smoke Abatement Society, in 1905. Today the term is applied to any objectionable mixture of air pollutants, as in the case of *Los Angeles smog*, q.v., which is of photochemical origin. Other terms in use are "smaze", a mixture of smoke and haze, and "smust", a mixture of smoke and dust.

Smoke. The visible product of incomplete combustion, consisting of minute carbonaceous particles mainly less than 1μ in size. Meetham in his *Atmospheric Pollution* (Pergamon Press, London and New York, 1956) has stated that after examination of 100,000 smoke particles it was found that half the individual smoke particles were smaller than about $0·075\mu$, but half the weight of smoke was in particles larger than about $0·51\mu$, and there was a definite tendency for smoke particles to stick together in chains perhaps 1μ in length. See **Black Smoke; Brown Smoke; Clean Air Legislation; Dark Smoke; Ringelmann Chart**.

Smoke Control Area. An area containing domestic, commercial and industrial premises declared by order to be a "smoke control area" by a local authority exercising its powers under Section 11 of the Clean Air Act, 1956. The general effect is to prohibit the emission of all smoke from chimneys in the area, subject to any exemptions in force. See **Clean Air Legislation**.

H 213

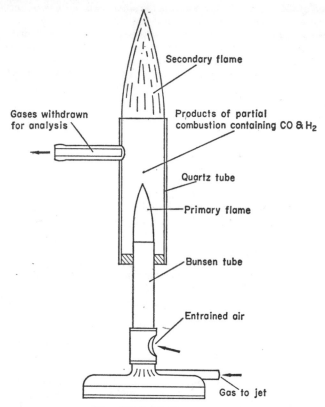

Fig. 25. *Smithells separator.*

Smoke Density Indicator. An instrument for measuring the opacity of smoke and particulate matter passing up a chimney. The principle is the projection of a beam of light of constant intensity across the interior of a stack. The beam of light, after crossing the stack, falls upon a *photoelectric cell*, q.v., producing an electric current whose intensity depends upon the amount of light falling upon it. This is used to operate an electrical indicating or recording instrument. The amount of smoke in the path of the beam of light determines the amount of light which falls on the photoelectric cell and the intensity of the current produced; this is used to indicate the density of the smoke. The instrument may be fitted with an alarm and/or a continuous recorder. See Fig. 26.

Smoke Eliminator Door. A special door for hand-fired natural draught Lancashire boilers developed by the British Fuel Research Station. It permits the introduction of additional volumes of secondary air after firing fresh fuel; it has openings of 32 in² for continuous air, and 29 in² for supplementary air.

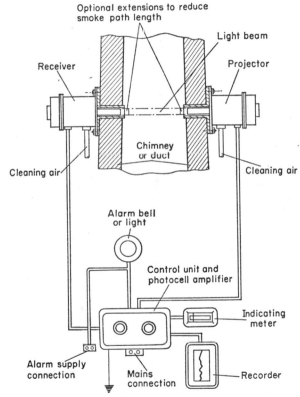

FIG. 26. *Diagrammatic arrangement of smoke-density indicator with meter recorder and alarm.*

215

Smoke Point Test. A test for the burning quality of kerosine. It indicates the height of flame which can be obtained without the formation of smoke, and thus provides a measure of the illumination likely to be obtained from a particular kerosine used in a particular type of wick-fed lamp.

Smokeless Zone. An area containing domestic, commercial and industrial premises declared by order to be a "smokeless zone" by a local authority exercising its powers under local legislation. The general effect is to prohibit the emission of all smoke from chimneys in the area, subject to any exemptions in force. See **Clean Air Legislation.**

Soaking Pit. An oil or gas fired furnace in which steel ingots are placed to provide an environment in which a uniform temperature throughout the ingot may be obtained; a uniform temperature ensures that the metal has the same degree of plasticity throughout the whole of the mass prior to rolling into sections or slabs.

Soda-base Grease. A grease with a high melting temperature; it is used in high speed bearings of an anti-friction type.

Sodium. Na. A metallic *element*, q.v.; at no. 11, at. wt. 22·99. It has a moderately low neutron-capture cross-section and can be used as a *coolant* in a *nuclear reactor*, qq.v.

Sodium Line Reversal Method. See **Line Reversal Method.**

Sodium Potassium Alloy. NaK. A low melting point alloy consisting of sodium and potassium; it may be used as a *coolant* in certain types of *nuclear reactor*, qq.v.

Sodium Sulphite. Na_2SO_3. An oxygen removing agent suitable for treating boiler feed water; it reacts with oxygen to give sodium sulphate. The reaction is somewhat slow and is often accelerated by adding a catalyst of cobalt salts. It adds appreciably to the dissolved solids in the boiler. See **Hydrazine.**

Soft Coal. *Coal*, q.v., relatively friable and subject to degradation on handling. See **Hard Coal.**

Softening Plant. Plant in which water is treated to remove *hardness*, q.v. See **Ion Exchange Process; Lime Soda Process.**

Solar Radiation. Energy radiated from the sun; only about half of the incoming solar energy reaches the earth's surface, much being reflected back into space by clouds and the remainder absorbed by the atmosphere. At the earth's surface the intensity of energy received may range from zero up to 300 to 350 Btu/ft^2 h. A device which enables the sun's rays to be effectively concentrated is known as a solar furnace; materials may be heated for long periods without contamination. The solar furnace is used as a tool in high-temperature research. Solar radiation has been used for the heating of buildings, steam generation, water heating, vaporization, distillation and drying. It has its best prospects of development in areas where sunshine is plentiful and conventional energy sources are scarce and expensive.

Solid Lubricants. Substances such as graphite or molybdenum disulphide which function as lubricants by reason of their crystal structure.

Solid Smokeless Fuels. See **Anthracite**; " **Cleanglow** "; " **Coalite** "; **Coke**; " **Gloco** "; " **Homefire** "; " **Phimax** "; " **Phurnacite** "; " **Rexco** "; " **Roomheat** ", " **Sebrite** ", " **Sunbrite** ", **Welsh Dry Steam Coal.**

Solute. A dissolved substance; hence solute + solvent = solution.

Solvent Spirits. White spirit and industrial solvents of various distillation ranges.

Soot Blower. A device used for cleaning boiler surfaces, using steam or air as the blowing medium. There are two main types of blower, (a) the single-jet or gun type, which is usually retractable; (b) the multi-jet type, which is usually non-retractable. Where deposits are likely to become hard, single nozzle blowers are essential, utilizing a pressure of 200 lb/in². Multi-jet blowers are more suitable for softer deposits with pressures varying from 80 to 150 lb/in².

Soot Blowing. A method of cleaning the external tube surfaces of a boiler while on load. Mechanical soot blowers are employed; these blowers may be operated by direct or remote manual control or by a fully automatic electric or compressed air sequential system. Either steam or compressed air may be used as the blowing medium. The use of suitable equipment may raise the efficiency of a shell boiler by 5 to 7 per cent for a steam consumption of less than 0·5 per cent of the total boiler output. Soot blowing should be carried out at least once a shift.

Sour. Having an unpleasant smell; positive to the *doctor test*, q.v. Sourness indicates the presence in gasolines, naphthas and refined oils of hydrogen sulphide and/or mercaptans. The term sour gas is applied to *natural gas*, q.v., containing more than 1·5 grains of *hydrogen sulphide*, q.v., per 100 ft.³

Source Control. The elimination, before or during ultimate consumption of potential air contaminants contained in raw materials, thus preventing the emission of contaminants to the atmosphere.

Source, Point, Line and Area. A point source is the plume from a single chimney; a line source is the emission from a row of industrial chimneys; an area source is a cluster of smoking chimneys in an urban residential or industrial neighbourhood.

Spark Arrester. A screen-like device situated on the top of a stack or chimney, or furnace exit, to reduce the amount of incandescent material expelled to the atmosphere.

Sparklers. White hot particles in a furnace; in a pulverized fuel-fired furnace this is an indication that oversized particles are passing the classifier.

Specific Gravity. (a) In respect of a substance, the density of that substance in relation to the density of water. Thus the specific gravity of water is taken as unity. The specific gravities of some common substances are:

Aluminium	2·6	Crude oil	0·8 to 0·97
Lead	13·6	Coal	1·2 to 1·7
Benzine	0·899	Pyrites	4·0 to 4·9
Bitumen	1·00 to 1·10	Gypsum	2·3

(b) Respecting a gas, the ratio of the density of that gas to the density of air. Thus, the specific gravity of air is taken as unity. The specific gravities of some common gases are:

Steam (at 212° F)	0·469
Hydrogen	0·0606
Oxygen	1·105

See **Apparent Specific Gravity; True Specific Gravity.**

Specific Heat. The amount of heat required to raise a unit mass of a substance through one degree, expressed as a ratio of the amount of heat required to raise an equal mass of water through the same range. If expressed on the Fahrenheit scale, the heat required is expressed in Btu; if on the Centigrade scale, then the heat required is expressed in Chu. Water has a specific heat of 1·0 at 60° F (15·5° C). Other substances have specific heats as follows: aluminium, 0·22; copper, 0·09; ice, 0·49; iron, 0·11; lead, 0·32; steam, 0·5 to 0·64. The specific heat of gases may be measured at constant volume or at constant pressure. For air the specific heat at constant volume is 0·17; at constant pressure, 0·24.

Specific Impulse. In rocket systems, the pounds force of thrust developed per unit mass of propellant per second.

Spectra. Phenomena observed when all the electromagnetic radiations characteristic of a particular source of radiant energy are separated into an array of constituent colours, wavelengths and frequencies; the wavelengths may be separated by refraction in a transparent prism, by diffraction in crystalline solids or by diffraction from a ruled grating. See **Spectrophotometry; Spectroscopic Units.**

Spectrophotometry. A technique of chemical analysis based on the absorption or attenuation by matter of electromagnetic radiation of a specified wavelength or frequency; the region of the electromagnetic spectrum most useful for chemical analysis is that falling between 2000 Å and 300μ, i.e. in the infra-red, visible and ultra-violet regions. A simple spectrophotometer consists of a source of radiation, such as a hydrogen or tungsten lamp; a monochromator containing a prism or grating which disperses the light so that only a limited wavelength or frequency range is allowed to irradiate the sample; and a detector, such as a photocell, which measures the amount of light transmitted by the sample. The absorbance of the sample is read directly from the measuring circuit of the spectrophotometer; when the absorbance of a sample is measured and plotted as a function of wavelength an "absorption spectrum" is obtained. Calibration or standard curves are prepared by measuring the absorption of known amounts of the absorbing material at the wavelength at which it absorbs most strongly. See **Chromatography; Spectra; Spectroscopic Units.**

Spectroscopic Units. Units of length used in stating the wavelengths of spectral lines in different parts of the spectrum; these include:

$$\text{micron } (\mu) = 10^{-4} \text{ cm}$$
$$\text{millimicron } (m\mu) = 10^{-7} \text{ cm}$$
$$\text{ångström } (\text{Å}) = 10^{-8} \text{ cm}$$
$$\text{X-unit } (\text{XU}) = 10^{-11} \text{ cm}$$

See **Spectra; Spectrophotometry.**

Spill Type Burner. A *wide range pressure jet burner*, q.v.; the velocity of oil through the tangential ports is maintained at high discharge rates by recirculating a controllable amount of the oil which enters the swirl chamber back to the suction of the pump. See Fig. 27.

Splint Coal. Coal consisting predominantly of *attritus*, q.v.

Spontaneous Combustion. Combustion which occurs of itself as a result of the combination of fuel with oxygen; this is liable to occur

219

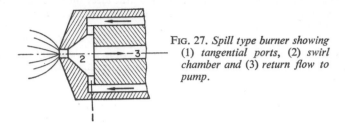

FIG. 27. *Spill type burner showing (1) tangential ports, (2) swirl chamber and (3) return flow to pump.*

when the ventilation of a coal stack or heap is not sufficient to carry away the heat liberated, or when the stack is not sealed off well enough to prevent air from entering it.

Spreader or Sprinkler Firing. A method of firing a boiler or furnace with solid fuel. In this method the coal is spread evenly and thinly over the entire fuel bed, but with rather less at the very back. The principle of this system is imitated by the automatic sprinkling stoker. It is claimed that this method produces better results and more steam than any other but it is the most liable to create smoke. When hand-firing, the tendency to create smoke can be partly ameliorated by adopting the well-known principles " little and often; level and bright ". The frequency of firing depends on the load but it should be regular, say every 10 minutes. The volatiles are thus given a reasonable chance of being burnt. It is only when the boiler is working at moderate load, however, that smoke-free operation is likely to be achieved. See **Firing by Hand; Sprinkler Stoker.**

Spring-loaded Safety Valve. A type of *safety valve*, q.v., in which the valve is kept in place by a spring; it has superseded the dead weight valve on high-pressure boilers. These valves may be single or double, with cast iron body for the lower pressures and cast or forged steel, or bronze, for higher pressures. An easing lever is fitted so that the valve can be lifted off its seat to make sure it is not sticking. In modern types, a cap is fitted over the adjusting screw and secured to the valve spindle by a cotter and padlock to prevent unauthorized interference.

Sprinkler Stoker. A *mechanical stoker*, q.v., which throws or sprinkles the fuel on to the grate, giving a thin and uniform fuel bed. As the volatiles are released from the top of the fuel bed, the mode of combustion is described as " over-feed ". The thin fuel bed tends to lessen the effects of caking properties and in this respect the stoker is flexible as regards type of fuel. It also greatly facilitates the combustion of the volatile matter given off by the coal. The chimney of a well-operated stoker should show only a slight haze, but the *grit carry forward*, q.v., is likely to be high when the fuel contains a considerable proportion of fines, particularly at high combustion rates. Coal is fed from a hopper, being pushed forward on to a feeder plate and projected into the furnace by either a rotary sprinkler or a spring-loaded shovel. Burning rates of up to 40 lb/ft² grate can be obtained.

Spun Glass. An insulating material, but not suitable for surfaces above 900° F (482° C).

Squish. A method of obtaining turbulence in the combustion chamber of an internal-combustion engine by using an irregular shape between the top of the piston and the cylinder head.

Stabilizer. A petroleum distillation plant in which "wild" or low-boiling hydrocarbons are removed under pressure from distillate or gasoline. The stabilized gasoline, with a boiling point between 86 and 158° F (30 and 70° C) may be used in blending low and medium grade motor spirits and aviation turbine gasolines.

Staffordshire Kiln. A top-fired transverse arch or chamber kiln used in the brick-making industry; it is fired by coal or oil.

Standard Temperature and Pressure (S.T.P.). A conventional reference condition of 0° C and 760 mm of mercury. It is synonymous with "normal temperature and pressure" or n.t.p.

States of Matter. The states in which matter can exist as a solid, liquid or gas. Solids are substances which tend to retain their shape and size indefinitely without deformation. Liquids are substances which flow to take up the shape of the vessel containing them. Gases also flow to take up the shape of the vessel containing them, but also expand to occupy all the available space. The state of a substance can be changed by a sufficient change in its heat content or its pressure.

Static Pressure. The pressure of a fluid in motion, or at rest, exerted perpendicularly to the direction of actual or possible flow, as distinct from *velocity pressure*, q.v. See **Total Pressure.**

Static Rectifier. A transformer-rectifier for converting the alternating current of normal electricity supply into high tension direct current electricity; this unit has found wide use in electrostatic precipitators. The use of selenium elements has superseded the original copper-oxide type; silicon types have been developed but are more expensive. The main advantages of the static rectifier are: (a) silent operation; (b) total enclosure in oil without any moving

H* 221

parts; (c) requires no radio and television suppression devices; (d) simple to operate from a remote position; (e) no nitrous oxide is generated. For performance under erratic or adverse process conditions they are not considered so adaptable as mechanical rectifiers See **Electrostatic Precipitator; Mechanical Rectifier; Rectifier.**

Static Stability. A fundamental concept in meteorology, it refers to what happens to a parcel of air after it has been given an initial vertical displacement, either upwards or downwards. If, after an upward displacement, the parcel is found to be warmer (less dense) than its surroundings its buoyancy will make it move farther from its original position; it will continue to move until its temperature (and density) becomes equal to that of its surroundings. In this instance in which a parcel moves farther from its original level, the surrounding atmosphere is said to be statically unstable (or just "unstable"). If after displacement, however, the parcel is found to be colder (more dense) than its surroundings, its buoyancy will tend to return the parcel to its original level; the surrounding atmosphere is then said to be statically stable (or just "stable"). In the intermediate stage when the vertical motion is neither encouraged nor opposed the atmosphere is said to be neutrally stable. The concept of static stability is of considerable importance in the study of the dispersal of air pollutants in the atmosphere. See **Lapse Rate.**

Stator. The stationary part of an *alternator*, q.v.

Steam. Generated from water, the most widely used heat transport fluid; it is cheap, non-toxic, offers ease of control and distribution, and has a high heat transport capacity. It contains about 25 times as much heat as the same weight of air at the same temperature.

Steam Blast Burner. A type of *oil burner*, q.v., consisting of a double concentric tube which allows steam at 20 to 60 lb/in² above the oil pressure to impinge on the oil feed and atomize it. See **Atomization.**

Steam Conditions. The pressure and temperature of steam, specified usually at the boiler main stop valve.

Steam Cycle. The cycle of events through which steam as a working fluid passes.

Steam Engine. A *heat engine*, q.v., in which the expansion of steam is utilized to effect the movement of a piston in a cylinder; in its simplest form the steam expands from the initial pressure to the exhaust pressure in a single stage. In a multiple-expansion engine, the expansion of steam is divided into two or more stages which are performed successively at falling pressures in cylinders of increasing size. In the compound engine, high and low pressure cylinders are used in series. In a triple-expansion engine, the steam expands successively in a high pressure, intermediate pressure, and low pressure cylinder, working on the same crankshaft.

Steam Flow/Air Flow Ratio. A relationship which has been employed as a means of combustion control in solid-fuel fired steam boilers. See **Combustion Meter**.

Steam Reforming. The treatment of hydrocarbons with steam in the presence of a catalyst to produce carbon monoxide and hydrogen.

Steam Tables. Published tables giving experimentally tested data for total heat, volume and entropy of steam at various temperatures and pressures. The Callendar Steam Tables are an example, being available in both Fahrenheit and Centigrade temperature scales and in complete and abridged forms.

Steam Trap. A device fitted at the lowest point of a steam pipework system to provide automatic *condensate*, q.v., recovery. Steam traps may be classified into three main groups: (a) mechanical group which, through the action of a float, open to condensate and close to steam; (b) thermostatic group, which open or close according to the temperature; (c) thermodynamic group, which work on the difference in velocity between steam and condensate flowing across a simple valve disc. Other types are also available.

Steam Turbine. A *heat engine*, q.v., in which jets of steam impinge upon blades attached to one or more discs or wheels mounted on a shaft supported in bearings. The shaft thus rotates and may be used to drive an electric generator or other machinery. The successful development of the steam turbine was largely due to Sir Charles Parsons (1854–1931). See **Back-pressure Turbine; Compound Turbine; Condensing Turbine; Gas Turbine; Impulse-type Turbine; Pass-out Turbine; Turbine Cylinder; Turbine Rotor; Turbine Steam Conditions; Turbine Steam Rate; Turbine Thermodynamic Efficiency; Turbo-Alternator.**

Steel. An alloy of iron and carbon. A typical analysis of mild steel, constituents being expressed in percentages, is: carbon, 0·2; silicon, 0·04; sulphur, 0·05; phosphorous, 0·05; manganese, 0·5; balance, iron. In alloy steels some of the iron is replaced by elements such as nickel, chromium and tungsten. Steel is used for boiler shells, steam and water drums, economizers, air pre-heaters, conveyors, hoppers, bunkers, tubes, pipes, ducts, and all structural work. Alloy steels are used to meet extreme conditions of heat and/or corrosion.

Steelmaking Furnace. See Ajax Furnace; Bessemer Converter; Electric Arc Furnace; Kaldo Furnace; Linz-Donawitz Converter; Oberhausen Rotor Furnace; Open-hearth Furnace.

Stefan-Boltzmann Formula. A formula for calculating the values for radiation from a solid body:

$$W = \delta \varepsilon T^4$$

where W = radiant energy per unit area per unit time, Btu/ft² h

δ = Stefan-Boltzmann constant, $1 \cdot 71 \times 10^{-9}$ Btu/ft² h

ε = emissivity of the surface, a dimensionless number between 0 and 1

T = absolute temperature, °R.

Steradian. Unit of solid angle; the solid angle which, having its vertex in the centre of a sphere, cuts off an area of the surface of the sphere equal to that of a square having sides of length equal to the radius of the sphere.

Still. A closed chamber, usually cylindrical, in which heat is applied to a substance to change it into vapour.

Stoichiometric. The calculated combining weights of chemical reactions or processes based on the laws of conservation of mass and energy and the chemical laws of combining elements.

Stoichiometric Air. See **Theoretical Air.**

Stokes. The unit of *kinematic viscosity*, in the *metric system*, qq.v. See **Centistokes.**

Stokes' Law. An equation by means of which the free-falling velocity attained by a particle under viscous flow conditions may be calculated:

$$v = \frac{d^2 g(\sigma - p)}{18\eta}$$

where v = free-falling velocity, cm/s

σ = density of particle, g/ml

p = density of fluid, g/ml

g = gravitational acceleration, 981 cm/s²

η = absolute viscosity, poise

d = Stokes' diameter of particle, cm.

Stokes' diameter is the "equivalent free-falling diameter" within the range of validity of Stokes' law, i.e. for which the *Reynolds number* q.v., is less than $0 \cdot 2$. The "equivalent free-falling diameter" is the velocity of fall of a particle through a still fluid at which the effective weight of the particle is balanced by the drag exerted by the fluid on the particle. (Reference British Standard 2955 : 1958.)

Storage Heater. A specially constructed block which can store heat, providing space heating at a low cost. It is heated by electricity at " off-peak " times, giving up the heat during the day.

Straight-Run Distillation. Continuous distillation which separates the products of petroleum in the order of their boiling points without cracking. Hence "straight-run gasoline" or gasolines as supplied by primary distillation without further treatment.

Straight-run Products. Products produced by straight-run distillation.

Strake. A cheap and simple method of preventing tall chimney stacks from oscillating in high winds developed by the National Physical Laboratory: the method consists of applying metal strips, called strakes, to the top section of a stack arranged in the form of a helix. Each strake consists of a plate attached edgewise to the surface of the cylinder. It is wound round the length of the cylinder in a rising spiral. The number of such strakes required is not critical but experiments carried out to determine the optimum arrangement of strakes led to the recommendation of a three-start configuration with strakes of 0·09 diameter high, wound with a helix pitch of five diameters. Structures or structural members of bluff section tend to oscillate in wind because of the vortices which are produced alternatively at the sides of the body and shed into the wake behind the body. This vortex formation gives rise to an alternating force in the cross-wind direction, and consequently the oscillatory displacements are produced in that direction. The strakes upset the correlation (i.e. the phasing) of the vortices shed from different parts along the length of the cylinder, and in this way a build-up of the forces tending to set up oscillations is prevented.

Stream Days. In respect of plant, *time efficiency*, q.v., expressed as operating days per year.

Strip Mining. See Open Cut Mining.

Sub-bituminous Coal. Coal intermediate between *lignite* and *bituminous coal*, qq.v., in appearance and properties.

Subcritical. The condition of a *nuclear reactor*, q.v., in which the rate of fissioning is not sufficient to sustain a chain reaction.

Sub-hydrous Coal. *Coal* containing less *hydrogen*, qq.v., than is normal for the type species. See **Orthohydrous Coal; Per-hydrous Coal.**

Subscribed Demand Tariff. A *tariff*, q.v. for the supply of electricity consisting of a unit charge and a charge related to the demand for which the consumer wishes to subscribe.

Subsidence Inversion. An *inversion*, q.v., formed usually well above the earth's surface as a result of the slow descent of air which has become warmer through *adiabatic*, q.v., compression.

Substation. Any premises or enclosure containing apparatus for transforming or converting electricity from one voltage to another, e.g. (a) stepping up to a high voltage for transmission, or (b) stepping down to a low voltage for distribution.

Suction Pyrometer. A device for measuring temperatures, when errors due to radiation from neighbouring surfaces are likely to occur; it comprises an open-ended tube containing a bare thermocouple surrounded by radiation shields. This assembly is inserted into a duct or flue and a sample of gas is drawn continuously through it. In a sonic suction pyrometer the gases are sucked through a nozzle at the speed of sound. In some cases suction pyrometers are water-cooled. See **Pyrometer; Thermo-Electric Pyrometer; Thermometer.**

Sulphur. S. A non-metallic *element*, q.v.; relative atomic weight 32·064. It occurs in nature in the free state and combined as sulphides and sulphates. It is present in coal and oil, being derived from the substances from which they were formed. The sulphur may be either chemically or physically mixed with the fuel. The sulphur content of British coals ranges from 0·5 to 3·5 per cent, with an average of 1·6 per cent. Sulphur is present in coal in three forms: (a) *pyrites*, q.v.; (b) organic sulphur compounds, cleaning processes having virtually no effect on this type of sulphur; (c) sulphates, present only in very small quantities and rarely exceeding 0·03 per cent. The ratio of pyritic to organic sulphur varies considerably. High *rank*, q.v., coals tend to be low in sulphur, medium rank coals high in sulphur, while low rank coals appear to be fairly average. During the combustion of coal, most of the sulphur is released to the atmosphere, mainly as *sulphur dioxide* together with a small amount (3 to 5 per cent) of *sulphur trioxide*, qq.v. With coal and coke burned in domestic heating appliances, about 20 per cent of the sulphur is retained in the ash or clinker. In industrial boilers and furnaces, only about 10 per cent of the sulphur is retained in the ash. The crude oils of the Middle East have an average sulphur content of about 2·5 per cent; most of this sulphur appears in the final products, being less than 1 per cent in diesel and gas oils and up to 4 per cent or more in residual oils.

During the combustion of oil virtually the whole of the sulphur is released to atmosphere, mainly as sulphur dioxide.

Sulphur Content, Statutory Limitations on. Legal limitations on the amount of sulphur allowed in fuels which are to be burned in certain areas; the purpose is to restrict the amount of *sulphur dioxide* in the atmosphere by *source control*, qq.v. In May, 1966, New York City modified existing Regulations setting a maximum of 2·2 per cent sulphur content in the fuel burned within the City by October, 1969, requiring instead that the standard be met by October, 1966, and that the standard be lowered to 2 per cent at the end of three years and 1 per cent at the end of five years.

Sulphur Dioxide, SO_2. A colourless, pungent gas formed when sulphur burns in air. It is considered to be one of the most important air pollutants; most of the sulphur dioxide in the general atmosphere comes from the combustion of the sulphur present in most fuels. The following are the average percentages of sulphur in fuels in common use in the United Kingdom: coal, 1·6; coke, 1·3; domestic fuel oil, 0·1; gas and diesel oil, 0·3 to 1·5; industrial fuel oil, 1·0 to 4·0; kerosine (paraffin), 0·03; coal tar fuels, 0·5 to 1·0; gas, 0·02. All the sulphur in oil, and from 80 to 90 per cent of that in coal and coke, is emitted from the chimney as sulphur dioxide, the remainder being retained in the ash.

Sulphur Trioxide. SO_3. A constituent of flue gases from *sulphur*, q.v., bearing fuels, frequently to the extent of 3 to 5 per cent of the *sulphur dioxide*, q.v., present. Several mechanisms appear to contribute to its formation; in every case a supply of oxygen is necessary. Thus the reduction of excess air tends to inhibit the formation of this corrosion-promoting gas. See **Acid Soot; Dew Point.**

"Sunbrite". A *hard coke*, q.v., carefully prepared and selected for domestic heating appliances such as room heaters, boilers, cookers, fires with under-floor primary air for combustion and fan assisted open fires. It is manufactured by the *National Coal Board*, q.v., the Steel Companies and other hard coke producers. See **Authorized Fuels; Smoke Control Area.**

Superadiabatic Lapse Rate. A *lapse rate*, q.v., greater than the dry adiabatic lapse rate, i.e. greater than 5·4° F (3° C) per thousand feet. The dry adiabatic lapse rate is often exceeded by a factor of several times near a land surface which is strongly heated by solar radiation. Turbulence in the atmosphere is strong and the dilution of waste industrial gases more rapid than in average or neutral conditions. See **Inversion.**

Supercritical Once-through Boiler

Supercritical Once-through Boiler. Known also as the Benson boiler, after its British inventor, a water-tube forced circulation boiler in which the feed water is heated, evaporated and superheated in a single passage, through a number of tubes in parallel. The original design was based on the principle that water at the critical pressure of about 3200 lb/in^2 needs no latent heat for conversion into steam. All the water flashes into steam when the necessary temperature is reached, so that no steam release surface and no steam drums are required. The use of small bore tubes and the absence of drums reduces the weight of the pressure parts and of the supporting structures. Britain's first commercial supercritical O.T. boiler was commissioned at the Margam "B" power station of the Steel Company of Wales Ltd. by Messrs. Simon Carves Ltd. The boiler, completed in 1960, was designed for steam conditions of 3300 lb/in^2 at 1060° F (571° C).

Supercritical Steam Pressure. Steam pressure at or above the critical pressure of 3200 lb/in^2 abs., at which pressure no latent heat is required to convert water into steam.

Supereconomic Boiler. A three-pass *Economic boiler*, q.v., in which the second and third passes of fire-tubes are situated below the main furnace tubes; this system ensures maximum heat transfer and imparts a vigorous circulation in what in other designs is a "dead water" area.

Super-grid. A high-voltage electricity transmission system operated by the *Central Electricity Generating Board*, q.v. It comprises a 275 kV (275,000 V) transmission network now being supplemented by a 400 kV (400,000 V) transmission system. See **Grid.**

Superheated Steam. Steam subjected to additional heating after leaving the boiler; the heat added to the steam is *sensible heat*, q.v. The *specific heat*, q.v., of steam varies between 0·45 and 0·65; consequently the addition of 1 Btu to 1 lb of steam raises its temperature by about 2° F, i.e. it produces 2° F of "superheat". A sufficient amount of superheat prevents harmful and wasteful condensation in steam turbines and steam engine cylinders. See **Superheater.**

Superheater. A *heat exchanger*, q.v., used in boilers to raise the temperature of the steam above that at which it leaves the boiler. The installation of a superheater leads to a reduction in steam requirements wherever it can be usefully employed, e.g. in supplying superheated steam to steam engines and turbines. Superheaters are constructed of small bore carbon and alloy steel tubes. In design, a superheater may consist of a series of U-shaped tubes connected to

228

headers, or a number of "hair pin" bends. In shell boilers the superheater is installed at the back end of the main furnace tubes where gas temperatures may range from 1000 to 1600 °F (538 to 871° C). Superheaters exposed to the radiant heat of the furnaces are described as "radiant superheaters"; those which derive heat solely from the hot flue gases are known as "convector superheaters". In the *water-tube boiler*, q.v., their description varies with position; if in the space over the water-tubes they are known as "over-deck", if between the water-tubes as "inter-tube", and if between banks of water-tubes as "inter-bank". See **Superheated Steam.**

Surface Combustion. *Combustion*, q.v., in the immediate vicinity of a hot surface; it has been found that hot surfaces have the property of increasing the rates of combustion of gases in air, temperatures of up to 3450° F (1900° C) being attainable.

Surface Condenser. A *condenser*, q.v., in which the cooling water flows through a large number of tubes, the steam condensing on the outer surfaces of the tubes.

Surface Ignition. In respect of an internal combustion engine, hot deposits in the cylinders which may ignite the cylinder charge at the wrong time.

Surface Moisture. Another name for *free moisture*, q.v.

Surface Tension. A characteristic of liquid surfaces whereby, due to unbalanced molecular cohesive forces near the surface, they appear to be covered by a thin elastic membrane in a state of tension. Surface tension is measured by the force acting across unit length in the surface, the force being expressed in dyn/cm. At 50° F (10° C), water has a surface tension of 74·22 dyn/cm, and benzene a surface tension of 30·22 dyn/cm.

Suspensoid. Suspended matter.

Suspension Firing. The firing of *pulverized fuel*, q.v., which burns in suspension. The common methods of firing are vertical, horizontal, tangential, cyclonic, and opposed inclined.

Suspension Gasifier. A gasifier in which powdered fuel is held in suspension by the gasification medium, e.g. air.

Sweet. Having a pleasant smell; negative to the *doctor test*, q.v. Sweetness indicates the absence in gasolines, naphthas and refined oils of hydrogen sulphide and mercaptans.

Sweetening. The process by which petroleum products are improved in odour and colour by oxidizing the sulphur products and unsaturated compounds.

Swelling Number. In respect of coal, the number of the standard profile most nearly corresponding to the coke button obtained under test, taking the average of five determinations; the property of swelling is determined by the crucible swelling test described in B.S. 1016: Part 12: 1959. The standard numbered profiles are from 1 to 9 in half units. See **Gray-King Assay.**

Swimming Pool Reactor. A *nuclear reactor* using water as *moderator* and *coolant*, qq.v., in the form of a tank of water with the fuel elements suspended well below the surface so that the water also acts as a shield. Such reactors are often used for the study of shielding problems.

Swivel Type Damper. A damper or draught regulator used, for example, with economizers. A lever is attached to the swivel and works in a quadrant with two holes in it so that the damper may be held firmly in position by a steel pin.

Synchronous Motor. A motor designed for alternating current, the speed of the motor being directly proportional to the frequency of the supply.

Synthesis. A reaction of simple substances to produce more complex ones, e.g. the production of hydrocarbons from carbon monoxide and hydrogen.

Synthesis Gas. A mixture of gases made specifically for use in a synthesis process, e.g. gas used in the manufacture of ammonia and other chemicals such as methanol. See **Ammonia Synthesis.**

T

Tail Gas. Residual gas left after recovering the desired products from the gas stream.

Tanbark. A bark residue remaining after bark has been used in tanning operations. It contains a high percentage of water (60 to 70 per cent) and has a low calorific value of between 2500 and 3000 Btu/lb (1385 and 1665 Chu/lb).

Tank Farm. Land on which a number of storage tanks are located.

Tank Gauge. A device to indicate the level of a liquid fuel in a storage tank. Tank gauges can be divided, in general, into two types: (a) floating indicator type, where a float is coupled to an external indicator which is marked to give the oil in the tank in gallons; (b) diving bell type, in which the oil in the tank exerts a pressure on the air in the bell causing an indicating fluid to stand at a level equivalent to that of the oil in the tank.

Tapping. Drawing off molten metal from a furnace, e.g. an *open-hearth furnace* or a foundry *cupola*, qq.v.

Tar Distillation. The fractional distillation of tar by the application of heat to produce *coal tar fuels*, q.v., and other products. Two types of still are in use: (a) batch still, perhaps hand-fired with coke for intermittent production; (b) pipe still, for continuous distillation. In one example, crude tar is heated in a pipe still in two stages—to effect dehydration and subsequent fractionation into light oil, carbolic oil, naphthalene oil, creosote oil, anthracene oil and pitch of medium hardness.

Tar Oil. A product obtained by the distillation of *coal tar*, q.v., comprising "creosote oil", "anthracene oil", and other substances. It has a calorific value of from 15,000 to 16,500 Btu/lb (8330 to 9165 Chu/lb). See **Coal Tar Fuels.**

Tariff. Method of charging for services, e.g. supplies of gas or electricity. See **Block Tariff; Bulk Supply Tariff; Flat Rate Tariff; Installed Load Tariff; Load/Rate Tariff; Maximum Demand Tariff; Restricted Hour Tariff; Seasonal Tariff; Subscribed Demand Tariff; Time of Day Tariff; Two-part Tariff.**

Teeming. The pouring of steel into ingot moulds, which yields an *ingot*, q.v., after the mould is stripped. The ingot is then "soaked" with heat in a *soaking pit*, q.v.

Temco Kiln. A thin-walled, steel-encased circular kiln, in use in the ceramic industry mainly for firing steel works refractories from 2012 to 2102° F (1100 to 1150° C), with either mechanical stokers or oil. The crown and walls, some 9 to 11 in. thick, are constructed of refractory insulation bricks. A kiln of similar type, fired by natural gas, is in use in the United States for firing ceramic glazed and unglazed sewer pipes and other building products.

Temperature. The degree of hotness or coldness of a substance measured in terms of a temperature scale. See **Temperature Scales.**

Temperature Entropy Diagram. A graph used when consideration is being given to the optimum efficiencies obtainable with power station steam cycles of various temperatures and pressures.

Temperature Scales. The scales in terms of which temperature may be expressed. The four in common use are: degrees Centigrade (Celsius) (° C); degrees Fahrenheit (° F); degrees Kelvin (° K) and degrees Rankine (° R). The last two are *absolute temperature*, q.v., scales. In 1948, the General Conference of Weights and Measures adopted the Celsius in place of the Centigrade scale; the scales are identical although the term "Centigrade" remains in common use.

Temperature Scales

See Centigrade (Celsius) Scale; Fahrenheit Scale; International Temperature Scale; Kelvin Scale; Rankine Scale.

Tempering. The adding of water to coal, usually as it reaches a bunker, to reduce the resistance of fines in the fuel bed to the passage of combustion air. Sufficient water is added to make the coal "ball" in the hand when squeezed, without wetting the hand. Some coals are difficult to wet with water alone, and for these exhaust steam is admitted at the bottom of the coal hopper or chute to complete the wetting.

Tertiary Air. A third supply of air sometimes introduced along the path of a flame after the *secondary air*, q.v., inlets, to assist with the complete combustion of long-flaming coals.

Tesla. Unit of magnetic flux density; the tesla is equal to one *weber*, q.v., per square metre of circuit area.

Tetraethyl Lead. See Gasoline Additives; Lead Susceptibility.

Tetrahydrothiophene. See Odorizer.

Theoretical Air. Or stoichiometric air. The amount of air required in theory to burn completely a given amount of fuel; it may be calculated from the chemical composition of the fuel. Theoretical air requirements per 10,000 Btu (5,555 Chu) gross for different fuels are on average: anthracite and coke, 7·75 lb; bituminous coals, 7·5 lb; petroleum oils, 7·45 lb. The weight of air required per pound of fuel may be calculated with sufficient accuracy from the formula:

$$W_a = 11·5\,C + 34·5\left(H - \frac{O}{8}\right) + 4·32\,S$$

where, W_a = theoretical amount of air, lb;

C = carbon content of fuel, per cent;

H = hydrogen content of fuel, per cent;

O = oxygen content of fuel, per cent;

S = sulphur content of fuel, per cent.

The values of C, H, O, and S are obtained directly from the *ultimate analysis*, q.v., for solid and liquid fuels; for gaseous fuels a conversion from the volumetric analysis may be carried out. One pound of air at *normal temperature and pressure*, q.v., occupies 12·39 ft³. The specific volumes of flue gases at N.T.P. are: (a) coal, 12·15 ft³/lb; and (b) oil, 12·3 ft³/lb. In practice additional or *excess air* q.v., is required to ensure complete combustion. It is a useful guide that approximately 10 lb of air is required for the complete combustion of any coal for every 10,000 Btu (5,555 Chu) *net calorific value*, q.v., with 33 per cent excess air. When carbon or hydrogen in any form is burned with the theoretical amount of air, the heat content per cubic foot of the combustion gases is virtually the same, that is 100 Btu/ft³ (55 Chu/ft³). Thus a bituminous coal with a net calorific value of 12,000 Btu/lb (6666 Chu/lb), burned with the theoretical amount of air necessary for complete combustion, will yield 120 ft³ of combustion gases, with a heat content of 100 Btu/ft³ (55 Chu/ft³). See **Excess Air.**

Theoretical Draught. The draught which is produced by the difference in static head of equal columns of atmospheric air and flue gas when at rest. The theoretical draught in a chimney can be calculated from the formula:

$$\text{Chimney draught (in. w.g.)} = 0 \cdot 256 \ HP \left(\frac{1}{T_1} - \frac{1}{T_2} \right)$$

where, $H =$ chimney height above breeching, ft
$P =$ barometric pressure, inHg
$T_1 =$ ambient temperature, ° R
$T_2 =$ average chimney temperature, ° R

To obtain the actual or natural draught, the losses due to friction must be deducted from the theoretical draught. See **Draught; Natural Draught.**

Therm. A unit of heat containing 100,000 British thermal units.

Thermal Conductivity. K. See **Conduction.**

Thermal Cracker. An oil refinery unit for the cracking of *hydrocarbons*, q.v., utilizing heat and pressure only; heavier straight-run products such as gas oils and residual stocks can be thermally cracked into lighter fractions by heating to about 950° F (510° C) at pressure within the range 300 to 600 lb/in². This type of cracking was the first developed, but the *gasoline*, q.v., produced by this method is inferior to that obtained when cracking takes place in the presence of a catalyst. See **Fluid Catalytic Cracking Unit (F.C.C.U.).**

Thermal Efficiency. The calorific content of the total useful heat or energy produced by a plant, expressed as a percentage of the calorific content of the total fuel consumed, or simply:

$$\frac{\text{Heat output}}{\text{Heat input}} \times 100$$

In respect of a boiler, "heat output" relates to the calorific value of the useful hot water or steam produced. In considering electricity generation, however, the overall thermal efficiency is the total calorific value of the electricity "sent out" expressed as a percentage of the calorific value (gross as fired) of the fuel used. This overall concept takes account of both boiler and steam turbine losses. In 1964, the average thermal efficiency of power stations owned by the Central Electricity Generating Board was 27·6 per cent, although the designed efficiency of the newest and largest stations was 37·5 per cent.

Thermal Fisson. Fission produced by *thermal neutrons*, q.v.

Thermal Liquid. A liquid used as a heat-transporting medium in the field of process heating and cooling. Thermal liquids include hot water, mercury, diphenyl-diphenyloxide (Dowtherm "A"), O-dichlorobenzene (Dowtherm "E"), molten salt mixtures and mineral oils. Dowtherm "A" and "E" are products of the Dow Chemical Company. Of the mineral oils, Mobiltherm 600 and Mobiltherm Light, produced by the Socony Mobil Oil Company, are aromatic mineral oils of lower viscosity than conventional mineral oils.

Thermal Neutrons. Neutrons in thermal equilibrium with the material in which they are moving.

Thermal Reactor. A *nuclear reactor*, q.v., in which the chain reaction is sustained primarily by fission brought about by thermal neutrons.

Thermal Reformer. An oil refinery unit for up-grading heavy gasoline, or naphtha, obtained as a fraction from the crude oil units. The naphtha is charged to a furnace where, at a temperature of about 1000° F (538° C) and a pressure of 100 atmospheres, the structures of the hydrocarbons are changed to yield a reformate having anti-knock properties much superior to those of the original naphtha. The reformate is chemically treated and, with light straight-run gasoline, then forms the basis of Regular grade gasoline.

Thermal Shield. A metallic shield placed around a *nuclear reactor*, q.v., between the core and the concrete *biological shield*, q.v. It is usually made of steel or other high density material and is intended to absorb some of the thermal neutrons, gamma radiation, etc. emitted by the core, thus reducing the energy dissipated in the concrete shield and protecting it from thermal damage.

Thermal Storage Boiler. Or Ritchie boiler. A boiler in which provision is made for meeting steam peaks or valleys by means of a substantial and controlled rise and fall of water level. The boiler shell has a diameter very much larger than in a normal boiler of similar heating surface. It operates at a pressure higher than that required in the factory, the process pressure being held constant by a reducing valve. A master firing gauge indicates the need for increasing or decreasing the rate of firing. The water level is allowed to rise during periods of low steam demand and to fall when meeting peak steam requirements. It is claimed that the thermal storage boiler enables peaks of 25 to 40 per cent above average steam demand, and valleys of similar size, to be met without any change in process steam pressure and without any change in firing rate.

Thermal Stresses. Stresses set up in heated metals.

Thermo Compressor. A compressor using high pressure steam to compress low pressure vapour.

Thermocouple. See **Thermo-electric Pyrometer.**

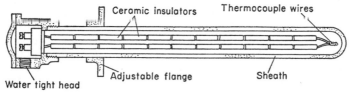

FIG. 28. *Construction of a thermocouple.*

Thermodynamics, Laws of. Fundamental statements regarding heat, energy and work. The First Law states that:

Heat supplied = work done + increase of internal energy.

The Second Law states that it is impossible to cause heat to pass from one body to another at a higher temperature without the aid of some external supply of energy.

Thermo-electric Pyrometer. A device for measuring temperature. It consists of two wires of dissimilar conductors joined together at one end—the "hot junction"—and enclosed in a protective sheath of steel, nickel chromium alloy or refractory materials. The two wires form a thermocouple. When the junction is heated an electromotive force is generated, the magnitude of which depends on the temperature difference between the "hot junction" and the cold ends of the conductors—the "cold junction"—as well as on the type of conductors used. The "cold junction" is connected to a measuring instrument. Base metal thermocouples are suitable for the temperature range $-328°$ F to $2012°$ F ($-200°$ C to $1100°$ C); rare metal thermocouples from $32°$ F to $2640°$ F ($0°$ C to $1450°$ C), generally. Thermocouples of the rhodium/platinum type can be used up towards the melting point of platinum $3216°$ F ($1769°$ C); some are capable of even higher temperatures. See Fig. 28. See **Pyrometer; Radiation Pyrometer; Suction Pyrometer; Thermometer.**

Thermometer. A device for measuring temperatures. See **Bimetallic Thermometer; Constant Volume Gas Thermometer; Liquid-in-Glass Thermometer; Liquid-in-Steel Thermometer; Resistance Thermometer; Vapour Pressure Thermometer.** Also **Line Reversal Method; Pyrometer.**

Thermometer Pocket. A metal or glass sheath to protect a thermometer from damage when it is inserted into a gas or liquid flow stream.

Thermopile, Flux-measuring. A device consisting of a number of thermocouples connected in series with alternate hot junctions coated with a neutron absorber such as boron; the current flowing is a measure of a neutron flux. This device has been used to investigate the variation of flux across a *nuclear reactor*, q.v.

Thermostat. A device designed to respond to temperature changes in a plant, operating controls either directly or indirectly should the temperature rise or fall above or below predetermined levels. Thermostats may be of the: (a) rod and tube type; (b) bi-metallic strip type; (c) liquid-expansion type; (d) thermocouple type. See **Pneumatic Controller.**

Thoron. A radioactive gas which is released from the soil into the atmosphere. Thoron gas decays before it has diffused more than a short distance above the ground.

Three-phase Power. An electrical system using three power-carrying wires, in comparison with "single-phase" power using two wires; by adding the third power-carrying wire, but using the same voltage and current, more power can be transmitted. Assuming a voltage of 250 volts and a current of 10 amperes:

Single-phase power $= 250 \times 10 = 2500$ W
Three-phase power $= 250 \times 10 \times \sqrt{3} = 4330$ W

Thus, in the above example, a three-phase system which has 50 per cent more wires than a single-phase system, achieves 73 per cent more power.

Three T's, The. Assuming a sufficient supply of oxygen has been directed to a fuel bed, the three essentials for efficient combustion—time, temperature and turbulence.

Throttling Calorimeter. Instrument for determining the *enthalpy*, q.v., and the quality of wet steam. Steam taken from a header in a sampling tube passes through a needle valve where it is throttled to a lower pressure; the temperature and pressure are taken in the low pressure calorimeter chamber. With the necessary readings a *Mollier chart*, q.v., is used to determine steam quality.

Throughput. The quantity per unit time of material undergoing a particular process or treatment.

Thylox process. A wet scrubbing process for the removal of *hydrogen sulphide*, q.v., from refinery and petroleum oil gas streams. The gas is scrubbed with a solution of ammonium thio-arsenate:

$$(NH_4)_3AsO_2S_2 + H_2S \rightarrow (NH_4)_3AsOS_3 + H_2O$$

See **Hydrogen Sulphide Removal.**

Time Efficiency. The fraction of the total time available during which a plant is in productive operation.

Time of Day Tariff. A *tariff*, q.v., for the supply of electricity which, in its simplest form, has a high rate in the day-time and a low rate at night, but a number of variations are possible.

Ton. The unit of measurement for the rate of refrigeration. A ton of refrigeration represents the rate at which heat is extracted to produce one ton of ice in 24 hours; this represents 12,000 Btu/h. The ton of ice refers to a short (U.S.) ton of 2000 lb.

Tonne. The metric ton equal to 1000 kilogrammes or 2204·62 lb. In Britain, the unit of mass mostly used when dealing with *nuclear reactor*, q.v., fuels.

Top Hat Kiln. An electric ceramic kiln in which the cover (sides and crown) is lowered over a base on which the ware is set for firing.

Topping. The distillation of crude oil to remove light fractions only.

Torbanite. See **Boghead Coal.**

Toroidal Combustion. A technique of combustion which uses the principle of the vortex to achieve continuous mixing of fuel and air. The action of the toroid arrangement is to increase the residence time of an atomized oil particle once it has left the burner. A toroidal system may be set up by a system of jets through which combustion air enters the combustion chamber. Toroidal combustion permits stoichiometric conditions to be maintained. See Fig. 29. See **Combustion; Stoichiometric.**

Total Gasification. Gasification without the production of residual coke.

Total Heat. Concerning steam, the sum of three components: (a) *sensible heat*; (b) *latent heat*; (c) *superheat*, qq.v., if any. The total heat in 1 lb steam at normal atmospheric pressure without superheat is:

Sensible + latent heat = total heat
180 + 971 = 1151 Btu
100 + 539 = 639 Chu

237

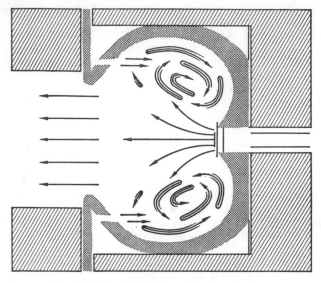

FIG. 29. *A single toroidal combustion chamber.*

The total heat of 1 lb steam at 100 lb/in² gauge and 430° F (221° C) is:

$$\text{Sensible} + \text{latent} + \text{superheat} = \text{total heat}$$
$$309 + 882 + 50 = 1241 \text{ Btu}$$
$$172 + 490 + 28 = 690 \text{ Chu}$$

As temperature and pressure rise, the amount of sensible heat increases, while the amount of latent heat decreases. At a pressure of 3200 lb/in² and temperature of 706° F (375° C) (no superheat), the "total heat" of 896 Btu/lb (498 Chu/lb) consists entirely of sensible heat. Also known as *enthalpy*, q.v.

Total Pressure. A pressure representing the sum of *static pressure* and *velocity pressure*, qq.v., at the point of measurement.

Total Radiation Pyrometer. A device for measuring temperatures, by measuring the intensity of all the wavelengths of the radiation emitted from a hot body. The heat rays radiated are focused by a mirror on to a sensitive thermocouple. The instrument is suitable for temperatures from 930° F (500° C) upwards. See **Radiation Pyrometer.**

Town Gas. Gas produced by a public utility for the general use of domestic, commercial and industrial consumers. In Britain much of this gas has been produced, until recently, almost entirely by the *carbonization*, q.v., of coal in gas retorts, the gas as supplied to consumers having a *calorific value*, q.v., of 450 to 550 Btu/ft³ (250 to 306 Chu/ft³). Town gas has been manufactured in the *horizontal retort*; the *intermittent vertical retort*; and the *continuous vertical retort*, qq.v. In recent years an increasing amount of town gas has come from new processes, such as the *Lurgi process*, q.v., and gas reforming processes using heavy and light oils and refinery *tail gases*, q.v. Supplies of gas have been further augmented by the importation of liquefied *natural gas*, q.v., from the Sahara. The current development of North Sea natural gas resources will result in a rapid displacement of conventional gasification processes.

Trace Heating. The heating of oil in a pipe by means of steam or electric heating elements.

Transformer. Apparatus for converting electrical energy received at one voltage to electrical energy sent out at a different voltage; in its simplest form it consists of two coils on a common soft iron or mild steel core. Alternating current in the " primary" winding sets up an alternating flux in the core, threading the "secondary" winding and generating an induced electromotive force across it. In a "step down" transformer, the output volts are less than the input volts; in a "step up" transformer, the secondary volts are higher than the primary volts.

Travelling Grate. A *mechanical stoker*, q.v., which differs from a *chain grate stoker*, q.v., only in relatively minor respects. The drive is independent of the grate, so that the expansion of the grate surface can be most easily allowed for and, in addition, individual grids of sections of the grate can be replaced without disturbing the drive. In a chain grate stoker, the drive is effected through the grate.

Trickle Feed Stoker. A *mechanical stoker*, q.v., to serve top-fired continuous brick kilns. It consists essentially of a hopper holding about 50 lb of small coal, at the bottom of which a rotating table discharges the coal in very small quantities at regular intervals through a feedspout to the kiln firehole. One stoker is required for each firehole, the mechanisms of each being operated by an electro-mechanical driving gear.

Trommel. A cylindrical screen with a centre shaft mounted nearly horizontally; the screen has round or square holes, and as coal or other material passes through it, the smalls pass through the holes.

Tropena. See **Bessemer Converter.**

True Specific Gravity. The ratio of the weight of a given volume of a sample of dried coal or coke, ground to pass through a 72-mesh B.S. test sieve, to the weight of an equal volume of water at the same atmospheric temperature. See **Apparent Specific Gravity; Porosity; Specific Gravity; Voidage.**

Tubular Precipitator. An *electrostatic precipitator*, q.v., consisting of vertical hexagonal tubes, arranged side by side through which the gas passes upward. The tubes may be up to some 10 in. in diameter and 12 ft in length and each contains an axial discharge electrode. While gas cannot by-pass the treatment zone, uniformity of gas distribution between the tubes is difficult to obtain; the re-entrainment of dust is high since it has to fall into the collecting hoppers through the incoming gas stream.

Tunnel Mixing Burner. A *gas burner*, q.v., in which the processes of mixing and combustion take place together in a refractory-lined quarl or tunnel. It provides a stable flame both for high and low flame speed gases, e.g. town and natural gas, over a wide range of conditions. Several variants of tunnel mixing burner designs are available.

Turbidity. A muddy condition of water from some natural sources, e.g. a river. If used in a boiler without prior treatment the suspended solids would quickly accumulate and produce a foam with the steam bubbles. All suspended solids should be filtered out before any other water treatment; if the solids are fine a coagulation process may be needed as well as filtration.

Turbine. See **Gas Turbine; Steam Turbine.**

Turbine Cylinder. The casing assembly of a steam turbine which houses the fixed blades and the rotor. See **Steam Turbine.**

Turbine Rotor. The rotating part within a *turbine cylinder*, q.v., to which the moving blades are attached. Rotors may be of the drum, disc, solid-forged or welded type.

Turbine Steam Conditions. The temperature and pressure of the steam at the turbine stop-valve. See **Steam Turbine.**

Turbine Steam Rate. The amount of steam, in pounds, consumed per kilowatt hour. See **Steam Turbine.**

Turbine Thermodynamic Efficiency. The ratio between the heat energy in the steam entering a turbine and the heat converted by the turbine into mechanical energy. See **Steam Turbine.**

Turbo-alternator. The combination of a *steam turbine*, q.v., and an electricity generating unit. In power stations, the turbine shaft is directly coupled to an *alternator*, q.v., which, when driven by the turbine, produces electrical energy. The world's most powerful generating unit (1000 MW) is at Ravenswood power station, New York; it is a cross-compound unit, i.e. with two lines of shafts. It was connected to the grid in June, 1965. The largest single-line shafts are to be found in the United Kingdom serving new 2000 MW power stations; these units have a capacity of 500 MW.

Turbo-generator. Synonymous with *turbo-alternator*, q.v.

Turbo-jet. A *gas turbine*, q.v., used for aircraft propulsion; the turbine drives the compressor and auxiliaries, the hot gases leaving the turbine discharge through the tail nozzle at a velocity of the order of 1700 ft/s. See **Turbo-prop.**

Turbo-prop. A *gas turbine*, q.v., used for aircraft propulsion, in which the energy of the hot gases is absorbed by a turbine to drive both the compressor and a propeller. See **Turbo-jet.**

Turbulence. The random movements of the air which are super-imposed upon the mean wind speed. An individual movement is called a " turbulent eddy"; it may have almost any size, and may move in any direction and at any speed.

Turn-Down Ratio. The ratio between full output and minimum output of an oil burner.

Tuyeres. Air ports which distribute air to a fuel bed.

Twin Fluid Atomizer. Another name for the *blast atomizer*, q.v., the term "fluid" being applied to both oil and air.

Two-colour Pyrometer. An instrument developed by the British Coal Utilization Research Association which measures the colour temperature of the individual particles in a gas stream. If the spectral emissivity of the coal particles does not vary substantially over the range of wavelengths viewed by the instrument, the colour temperature is little different from the true temperature. For carbon at 2000° K the colour temperature is known to be only 16 degK less than the true temperature. The colour temperature is determined by measuring the total amount of energy in each of two different narrow wavelength bands. The principle of the method is shown in Fig. 30. Images of particles moving in the furnace are focused on to a 300-micron aperture. The light from these images is then divided by a semi-silvered mirror so as to fall on two photo-multipliers of different spectral response characteristics. The output pulses from the two photomultipliers are displayed on a double-beam oscilloscope. The ratio of the pulse heights gives the colour temperatures of the individual particles in the gas stream. The range of the instrument is from 1742° F to 3632° F (950° C to 2000° C) with an accuracy estimated to be ±72 degF (40 degC) at 3632° F (2000° C). See **Colour/Temperature Scale; Pyrometer.**

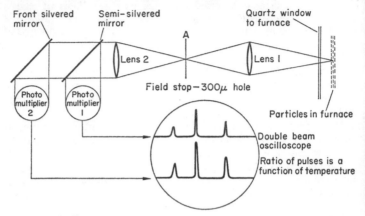

Fig. 30. *Optical system for two-colour pyrometer measuring colour temperature.*

Two-part Tariff. A composite charge for gas or electricity comprising a fixed charge per accounting period and a charge per therm or unit of gas or electricity consumed. The underlying principle is that the fixed minimum charge should cover roughly the fixed or overhead costs of production and distribution, and the charge varying with consumption should cover the variable costs of production. See **Tariff.**

U

Ultimate Analysis. A form of analysis which divides up a fuel into the percentages of carbon, hydrogen, oxygen, nitrogen, sulphur and ash which it contains; as a reminder of these constituents students sometimes use the mnemonic "no cash". See **Proximate Analysis.**

Ultrasonic Agglomerator. A device to agglomerate particles in suspension, thus facilitating their subsequent removal from a waste gas stream. Gases enter an agglomerating tower and are subjected to high frequency sonic vibrations; the fine dust, converted into larger aggregates, passes to a *cyclone*, q.v., or other device for collection.

Underfeed Combustion. *Combustion* in which *ignition*, qq.v., takes place from the top of the fuel bed, the *ignition plane*, q.v., travelling downwards against the air flow. The principle is illustrated in Fig. 31.

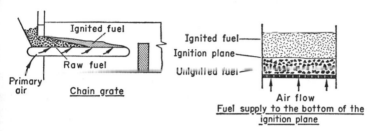

FIG. 31. *Underfeed combustion as illustrated in the chain grate stoker.*

This form of combustion is virtually smoke-free; it does not tend to form large coke masses and is relatively insensitive to coal *rank*, q.v. It occurs in the *underfeed stoker*, and in the *chain grate stoker*, qq.v. See **Overfeed Combustion.**

Underfeed Stoker. A *mechanical stoker*, q.v., in which fuel is supplied by an Archimedean screw, or by a reciprocating ram, into the bottom of a retort below the fire and is gradually forced up into the combustion zone. Combustion air is injected through ports or tuyeres into the fuel bed just below the combustion level. As the coal rises in the retort the volatiles released pass upwards through the burning fuel, mixing with the incoming combustion air, and burning at the top of the fire with a short flame. In correct conditions, when the fuel reaches the surface of the fuel bed only coke remains. This coke burns as it moves outwards towards the perimeter of the fire; combustion completed ash accumulates around the retort and must be removed by hand at intervals. All the air required for combustion is delivered through the tuyeres; it is not normally necessary to admit secondary air through the fire-door. The underfeed stoker is fitted to horizontal, vertical, sectional and central heating boilers. Where there is intermittent operation smoke may be emitted at times of stopping and starting, as the correct combustion conditions tend to be disturbed; this characteristic has been largely overcome by the development of overfire air jets and other techniques to maintain efficient combustion at all times. See **Black Centre Burning.**

Underground Gasification. The gasification of coal while still in the seam.

Underground Storage. The storage of gas in underground reservoirs and natural strata, instead of in gasholders above ground.

Unit of Electricity or Kilowatt-hour (kWh). The unit of electrical energy being the amount of energy corresponding to a power of one kilowatt (1000 watts) sustained for one hour. One kilowatt-hour $= 3412$ Btu at 100 per cent efficiency.

Unit System. A system for supplying *pulverized fuel*, q.v., to boilers or furnaces. One pulverizer supplies fuel direct to one or two furnaces. Unlike the *bin and feeder system*, q.v., this method does not require storage space or bins. It is the most popular system of P.F. firing.

Up-draught Kiln. Or bottle kiln. A ceramic kiln in which the gases pass through openings in the crown of the kiln; the kiln is surmounted by a superstructure in the shape of a bottle.

Uranium. U. The heaviest element occurring in significant quantities in nature; at. no. 92, at. wt. $238 \cdot 03$. Natural uranium contains 1 part of U^{235} to 139 parts of U^{238}. Uranium has a typically metallic appearance, being bright, hard and heavy. It is half as heavy again as lead, 1 in^3 weighing 11 ozs. See **Uranium Enrichment.**

Uranium Enrichment. The enrichment of purified uranium, derived from irradiated fuel elements, with the fissionable U^{235} isotope which was partly burned-up during exposure in the reactor. This is carried out in a diffusion plant at Capenhurst, Cheshire, England, where the uranium is exposed to metallic membranes which, as it were, filter off some of the non-fissionable U^{238} leaving a product enriched in fissionable U^{235} and suitable for re-use as a reactor fuel. See **Windscale Chemical Processing Plant.**

Urquhart Oil Gasifier. A high duty gasification unit in which fuel oil is reacted with less than the stoichiometric quantity of air to produce a gas with little or no carbon deposition; the reactions can be controlled to provide a range of gases varying in composition and temperature. See **Stoichiometric.**

U-tube Manometer. See **Draught Gauge.**

U-tube Viscometer. An instrument for measuring the *kinematic viscosity*, q.v., of fluids.

Vacuum Distillation. Distillation under reduced pressure. The effect is to reduce boiling temperature sufficiently to prevent decomposition or cracking of the material being distilled.

Vacuum Jets. Steam ejectors for removing air and non-condensable gases from barometric condensers on distillation equipment.

Valency. In respect of an *element*, q.v., the number of atoms of hydrogen that one atom of the element can combine with or replace.

Vanadium, V, A metallic *element*, q.v.; at. no. 23, at. wt. 50·94?; it occurs as an impurity in fuel oil to the extent of about 0·003 per cent.

Vane Control. The control of fan output by a system of adjustable vanes at the air inlet.

Vaporizing Burners. Or pot burners. Oil burners used in conjunction with domestic central heating boilers; the oil is vaporized by heat, the vapour mixing with the correct amount of air to achieve efficient combustion. There are two types of burner: (a) natural draught, utilizing a chimney of sufficient height to obtain the necessary draught; (b) assisted draught in which air is supplied by a fan at 1 to 3 in. w.g. Vaporizing burners are designed for light grades of fuel oil.

Vaporizing Oil. A fuel used in tractor engines of low compression ratio, i.e. about 4·5 to 1. The boiling range is similar to that of kerosine and it has an *octane number* q.v., of about 50. The fuel lacks readily volatile constituents and it is necessary to start tractor engines with motor gasoline before switching over to vaporizing oil. Also known as power kerosine.

Vapour. A gas at a temperature below its critical temperature; the latter being that above which the gas cannot be liquefied. Below the critical temperature a vapour can be liquefied by a sufficient increase in pressure, and cooling is unnecessary. Otherwise a gas must be cooled below its critical temperature and sufficiently compressed before it will liquefy.

Vapour Lock. A term applied to the condition when fuel starvation occurs in an engine, due either to the low fuel *volatility tolerance*, q.v., of the engine fuel system or to the excess volatility of the fuel itself. Vapour lock manifests itself either by hesitations during acceleration, or by stalling of the engine.

Vapour Pressure. The pressure exerted by a vapour, either by itself or in a mixture of gases; saturated vapour pressure means the pressure exerted by a vapour in contact with its liquid form. Saturated vapour pressure increases with rise of temperature.

Vapour Pressure Thermometer. A device for measuring temperature. It comprises a bulb partly filled with a volatile liquid such as methyl chloride or toluene and a pressure measuring element, the two being connected by a capillary tube. A change in temperature causes a change in pressure of the saturated vapour above the liquid. The instrument is suitable for a temperature range $-4°$ F to $+662°$ F ($-20°$ C to $+350°$ C). See **Pyrometer; Thermometer.**

Variable Speed Electric Motor. Electric motor the speed of which can be regulated; there are three types in common use, the a.c. slip-ring motor, the a.c. commutator motor and the d.c. motor. They have either shunt regulation or variable voltage control.

Vel. A unit of free falling speed. A particle has a free falling speed of one vel if it falls 1 cm per second in still air at 68° F (20° C). To

TABLE 9

Vel grading	Microns ($m \times 10^{-6}$) (Density 2 g/c³)
40	89
20	59
10	41
6	31
4	25
2	18
1	13
0·6	10
0·4	8

convert from feet per second to vels, multiply by 30·48. Table 9 gives the relationship between vels and particle sizes in microns, assuming a particle density of 2 grammes per cubic centimetre.

Velocity Pressure. Pressure caused by and related to the velocity of the flow of fluid, as distinct from *static pressure*, q.v.; a measure of the kinetic energy of the fluid. See **Total Pressure.**

246

Venturi Pneumatic Pyrometer. A *pyrometer*, q.v., based on the principle that if a quantity of hot gas is drawn through a venturi restriction, then cooled and passed through a similar restriction, the ratio of the differential pressures produced is inversely proportional to the ratio of the densities of the gas at the two points. If the gas obeys the *gas laws*, q.v., then the ratio is proportional to the ratio of the temperatures of the gas, measured from *absolute zero*, q.v. If the temperature of the gas is measured at the "cold" venturi, the temperature of the hot gas may be calculated:

$$T_h^\circ \text{ K} = \text{K} \frac{\Delta P_h}{\Delta P_c} T_c^\circ \text{ K}$$

where $T_h^\circ \text{ K}$ = absolute temperature of gas at "hot" venturi
  $T_c^\circ \text{ K}$ = absolute temperature of gas at "cold" venturi
  ΔP_h = pressure at "hot" venturi
  ΔP_c = pressure at "cold" venturi;
  K = a constant

The instrument may be used for gas temperatures above about 2550° F (1400° C); it responds to changes in less than two seconds.

Venturi Scrubber. A device for scrubbing industrial waste gases to remove dust before they pass to atmosphere. In a typical example, the gas to be cleaned flows into a venturi nozzle placed just before a point at which water is injected. The velocity of the gas may be raised to 300 ft/s or higher, this high gas velocity atomizing the water. The dust in the gas, now thoroughly wet, enters a second section where more water is added; the dust is removed from the gas by cyclonic action. This technique is highly efficient, figures as high as 99·8 per cent being quoted. The pressure drop is relatively high ranging up to 25 to 30 in. w.g. Power consumption is also high; water consumption is of the order of 5 to 6 gal/1000 ft³ of gas cleaned.

Venturi Tube. A tube which narrows to a throat and then gradually increases to the original diameter. It may be used as a device for measuring the flow of a gas or liquid, or with spray injectors as a means of scrubbing gases. See **Venturi Scrubber.**

Vertical Boiler. A steam raising plant which in its simplest form consists of a cylindrical vertical shell surrounding a fire box (combustion chamber) in the bottom of which is a grate on which the fuel is burnt. The fire box may contain cross-tubes to assist circulation.

247

Some types incorporate one or two banks of "smoke" tubes which increase the effective heating surface of the boiler. The working pressure is normally up to 150 lb/in². See **Boiler.**

Vibrating Screen. A screen used for removing the smaller sizes, say $<\frac{1}{2}$ in., from washed coal before sending it to market. The screen is electrically vibrated at a rate ranging up to 3000 vibrations per minute, with an amplitude of about $\frac{1}{8}$ in.

Virgin Stock. Oil derived directly from crude oil; "straight-run" stock. See **Straight-run Distillation.**

Viscometer. An instrument for measuring the *viscosity*, q.v., of a liquid. See **Ostwald Viscometer; Redwood Viscometer; Saybolt Furol Viscometer; Saybolt Universal Viscometer; U-tube Viscometer.**

Viscosity. The ease with which a liquid will flow. It is determined by recording the time required for a measured volume of liquid at a specified temperature to flow through an orifice of prescribed dimensions. Different methods are employed in various countries, those most commonly used being Redwood No. 1 and No. 2 (British), the results being expressed in "Redwood seconds"; Engler (Continental European); Saybolt Universal and Saybolt Furol (American). Results are frequently expressed in centistokes. See **Dynamic Viscosity; Kinematic Viscosity; Viscometer.**

Viscosity Breaking. The lowering or "breaking" of the viscosity of residuum by thermal cracking; this technique may be necessary to prepare a *residuum*, q.v., for fuel oil blending; the necessary conditions for cracking are about 900° F (482° C) and 250 lb/in².

Viscosity Index. An index which compares the change in the viscosity of a fluid obtained over the temperature range 100° F to 210° F (38° C to 99° C) with the change obtained over the same temperature range with a Pennsylvanian oil (of high viscosity index, good viscosity temperature characteristic) and a Gulf Coast oil (of low viscosity index, poor viscosity temperature characteristic).

Viscosity Index Improver. A lubricating oil additive which is intended to minimize changes of viscosity with temperature; typical compounds are polymeric hydrocarbons (isobutene) or esters (fumarate, methacrylate).

Viscous Oil Filter. An *air filter*, q.v., consisting of closely-spaced corrugated metal plates which are "wetted" with a special oil. Impurities in the air adhere to the oil and are removed by the continuous flushing of the plates with fresh oil. The resulting sludge is removed from the oil automatically. Collecting efficiencies of 85 to 90 per cent are claimed for 5-micron particles.

Vitrinite. A substance in coal derived from partly decayed bark and woody tissue completely impregnated by initially liquid decomposition products; a major constituent of *clarain*, q.v., in which it occurs in uniformly brilliant black layers.

V.L.N. Steelmaking Process. See Bessemer Converter.

V-notch Meter. A device for measuring the supply of feed water to a boiler. It can only be employed on the suction side of the feed pump. One type consists of a tank divided into three parts by means of a vertical baffle plate and a water-tight partition which carries a gun-metal V-notch plate or weir. The flow through the V-notch is automatically controlled in accordance with the demands of the feed pump by means of an equilibrium valve actuated by lever and float in the catch-box or hotwell end of the tank. The instrument is operated by a float which rises and falls with the level of the water in the V-notch and may be of the indicating, recording or counting type, or a combination of all three. See **Feed-water Meter.**

Voidage. The fraction of voids in a bed of coal or coke as calculated from the following expression:

$$1 - \frac{\text{bulk density (lb/ft}^3)}{62 \cdot 5 \times \textit{apparent specific gravity, q.v.}}$$

See **Porosity.**

Volatile Matter. The percentage loss in weight resulting from the heating of one gramme of coal under test conditions in a crucible from which air is excluded; moisture is excluded from this calculation. Volatile matter consists of a complex mixture of tarry vapours and combustible gases such as hydrogen, methane, ethane, benzene, etc. The volatile matter content of various solid fuels, expressed as a percentage of total weight, is as follows: (a) coke, 1 to 2; (b) anthracite, 4 to 8; (c) Welsh steam coal, 8 to 16; (d) caking coal, 20 to 34; (d) free burning coal, 34 to 40. Practically the whole of a liquid fuel consists of volatile matter. See **Volatility.**

Volatility. The ease with which a liquid fuel vaporizes. Tests on gasoline show the volume of vapour formed per unit volume of liquid at a series of different temperatures and indicate the tendency of a fuel to vapour locking. See **Vapour Lock; Volatility Tolerance.**

Volatility Tolerance. The ability of an engine to cope with excess volatility in a fuel; the volatility tolerance of an engine is interpreted in terms of the ambient temperature at which *vapour lock*, q.v., would occur on a standard motor fuel. Major factors which determine the volatility tolerance of a vehicle are the temperatures of the fuel system components, the most critical of which is usually the fuel pump, and the vapour handling capacity of the system.

Volt. The unit of potential difference and electromotive force; it is the difference of electric potential between two points of a conducting wire carrying a constant current of 1 *ampere*, q.v., when the power dissipated between these points is equal to 1 *watt*, q.v. Named after the famous Italian scientist, Volta (1745–1827).

Volumetric SO₂ Apparatus. An instrument for measuring sulphur dioxide concentrations in the general atmosphere; air is bubbled through dilute hydrogen peroxide, the sulphur dioxide being oxidized to sulphuric acid. Measurement is by titration with alkali. The instrument usually incorporates a smoke filter. Readings are usually taken every twenty-four hours. See **Lead Peroxide Candle; National Survey of Air Pollution, British.**

W

Wagon Tippler. A cradle in which a wagon of coal is secured; the cradle and wagon are rotated to empty the coal into a conveyor.

Wall-fired Furnace. A *water-tube boiler*, q.v., in which an array of pulverized-fuel burners is situated low on one wall; a typical array may consist of three rows each of six burners. The burners may be short-flame turbulent burners producing parallel coaxial annular swirling jets. The inner annulus feeds primary air and coal while the outer annulus feeds secondary air. Through the very centre of the burner, tertiary air may be admitted. The outlet from the furnace is at the roof; a projection from the rear wall of widely spaced water-tubes, termed a nose, assists in protecting the superheater tubes by partially chilling and partially deflecting the hot gases. The burners, instead of being all on one wall are sometimes arranged on opposite walls. The opposing jets produce a greater degree of turbulence in the furnace, compared with the one-wall arrangement. Front-wall-fired furnaces are used with finely ground high-volatile coals. See Fig. 32.

Washability Curve. A curve obtained by plotting the results of a *float and sink test*, q.v., from which the theoretical yield of floats and sinks may be read off.

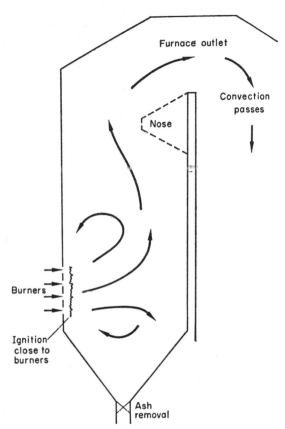

Fig. 32. *General shape of chamber and flow pattern in wall-fired water-tube boiler.*

Washed Coal. *Coal*, q.v., that has undergone a wet cleaning process to improve its quality.

Washery. A *coal*, q.v., preparation plant using wet cleaning processes. See **Coal Cleaning.**

Waste-heat Boiler. A boiler possessing a multi-fire tube system depending mainly or entirely on waste gases as a source of heat.

Water. The working fluid of steam systems, and a heating or cooling medium. Constituents of water which may give rise to difficulties in boilers are: (a) corrosive substances such as acid solutions and *dissolved gases*, q.v. including carbon dioxide, oxygen, hydrogen sulphide and ammonia; (b) scale-forming substances such as salts of calcium and magnesium; (c) foam-producing substances such as oil and the products of decomposition of sewage and humic matter. Water is used in nuclear reactors both as a *moderator* and *coolant*, qq.v. Ordinary or light water (H_2O) contains ordinary or light hydrogen (H); heavy water (D_2O) contains mainly heavy hydrogen or deuterium ($_1H^2$ or D).

Water Cooling. The recovery of water for reuse in a process. Besides reducing the basic cost of water there is also a saving in water treatment processes where these are necessary. There are several ways of cooling water so that it may be recirculated; these include (a) blast coolers ressembling a car radiator with mechanical draught; (b) water/water heat exchangers in which tubes containing the hot water are cooled by an external flow of cold water, the temperature of which is raised; (c) evaporative cooling in which hot water is brought into contact with the atmosphere. Evaporative cooling is the most widely used today.

Water Equivalent. In relation to a body, the weight of water which would require the same amount of heat to raise the temperature by one degree:

Water equivalent = weight of body × specific heat of body.

Water Gas. Gas produced by the action of steam on red hot carbon. The reaction is an *endothermic reaction*, q.v.:

$$C + H_2O \text{ (steam)} \rightarrow CO + H_2 - 4343 \text{ Btu (2413 Chu) lb/carbon}$$

The calorific value of the gas averages 300 Btu (167 Chu) ft³. It burns with a non-luminous bluish flame; often called "blue water gas". The manufacturing process is cyclic. Coke in a large cylindrical generator is blown with air to raise its temperature; this is followed by a "run" with steam. "Blow" gases are discharged to atmosphere. A typical analysis, the constituents being expressed in percentages is as follows: hydrogen, 48; carbon monoxide, 42; carbon dioxide, 4·5; nitrogen, 5·0; methane, 0·5. See **Carburetted Water Gas.**

Water Gas Shift Reaction. A reaction in which carbon monoxide and steam are partially converted to carbon dioxide and hydrogen. The reaction is:

$$CO + H_2O \rightarrow CO_2 + H_2$$

which proceeds to equilibrium only and not to completion. This reaction occurs in gasifiers at temperatures above about 1300 °F (700° C) influencing the composition of the final gas. It is also carried out over a catalyst at about 750 °F (400° C) to increase the H_2/CO ratio in gas, e.g. in the manufacture of hydrogen from water gas.

Water Gauge. A device to indicate the level of water in a boiler. All boilers with an evaporative capacity exceeding 300 lb steam/h should have two water gauges fitted at drum level, suitably protected by means of a shield of specially toughened glass. In large boilers, operating at pressures exceeding 900 lb/in² the water gauges may be replaced by two independent remote level indicators of the compensated manometric type. Water gauges must be so arranged that the lowest visible part of the gauge is not lower than the lowest safe working level of the water in the boiler. The gauges should be fitted with isolation cocks, together with a drain cock or valve with a discharge pipe. Water gauges should be tested regularly, a set procedure being followed by the stoker at the beginning of each shift. With large water-tube boilers, the water gauges are situated far above the firing floor. Illuminated indicators are often fitted so as to enable the gauges to be read from the firing floor.

Water Lancing. A process complementary to *soot blowing*, q.v., the use of a water jet to remove deposits from boiler surfaces. Often deposits with the greatest resistance to soot blowing are those most readily removed by water lancing.

Water Smoking. Or steaming, the release of water vapour during the period when ceramic ware is being dried by passing large volumes of air, at temperatures up to 480° F (250° C) through the setting.

Water Treatment. Measures to condition water so that it is suitable for use in boilers; the main aims of water treatment are: (a) to "soften" the water, i.e. to remove or neutralize the scale-forming salts; (b) to achieve the correct alkaline condition in the boiler water; (c) the removal of excess oxygen and carbon dioxide from the water. By adopting the correct methods to achieve these aims scale and deposits, corrosion and embrittlement, can be largely avoided and the steam supply protected against impurities. See **Carboxylic Resin; Caustic Embrittlement; Chelating Agents; Hydrazine; Ion Exchange Process; Lime-Soda Process; Sodium Sulphite.**

Water-tube Boiler. A steam generator consisting of a large number of closely spaced water-tubes connected to one or more drums which act as water pockets and steam separators. Draught fans, firing

I* 253

equipment, superheaters, economizers and air pre-heaters complete the installation. High rates of evaporation are possible, up to 4 million lb steam/h, at pressures up to 3500 lb/in². Thermal efficiencies of the order of 90 per cent are attained. Water-tube boilers are universally used for power stations, most of them being fired by pulverized fuel. See **Boiler; La Mont Boiler; Supercritical Once-through Boiler.**

Watt. The smaller unit of electric power; it is the power which gives rise to the production of energy at the rate of 1 *joule*, q.v., per second. With *direct current*, q.v., volts × amperes = watts. One kilowatt = 1000 watts. Named after the famous Scottish engineer, James Watt. See **Kilowatt**

Weakly Caking Coal. *Coal*, q.v., which does not become sufficiently plastic during carbonization to form a mechanically strong coke.

Weathered Coal. *Coal*, q.v., the character of which has changed due to the combined effects of chemical and physical action on exposure to weathering processes.

Weathered Crude. Crude petroleum which, during storage and handling, has lost an appreciable quantity of the more volatile components due to natural causes.

Weber. A unit of magnetic flux; it is the magnetic flux which, linking a circuit of one turn, produces in it an electromotive force of one *volt*, q.v., as it is reduced to zero at a uniform rate in one second.

Welsh Dry Steam Coal. A high *rank*, q.v., coal containing between 9 and 19·5 per cent *volatile matter*, q.v.; it is therefore somewhat more reactive than *anthracite*, q.v. As the name implies it is a product of the South Wales coal-fields.

Wet-back Economic Boiler. See **Economic Boiler.**

Wet-bottom Furnace. A furnace in which the ash is deposited and leaves the furnace in a molten condition. From 30 to 50 per cent of the ash leaves through the ash hopper. Also known as a slag-tap furnace. See **Dry-bottom Furnace.**

Wet Gas. A petroleum gas containing a relatively high proportion of hydrocarbons recoverable as liquids.

Wet Gas Holder. An inverted metal dome floating in a water tank with inlet and outlet gas connectors; the pressure on the gas is controlled by a system of balancing weights. See **Piston Type Gas Holder.**

Wet Gas Meter. A gas meter consisting of a casing in which a hollow measuring drum rotates in water. The drum consists of several gas compartments of equal volume; gas enters one compartment causing the drum to revolve, this gas being discharged while the next compartment is filling. The drum is connected to dials indicating accurately the volume of gas passed through. See **Dry Gas Meter.**

Wet Saturated Steam. Steam containing entrained water droplets. See **Dryness Fraction; Dry Saturated Steam.**

Wet Washer. Or scrubber, a device for removing particulate matter from gas streams. Wet washers or scrubbers fall into two main categories: (a) water film types in which the gases are brought into contact with water-covered surfaces so that the dust particles are retained and discharged in the water; (b) water spray types in which the gases are sprayed with water in a finely divided state. See **Dust Arrester; Venturi Scrubber.**

White Cast-iron. Hard *cast-iron*, q.v., which presents a silvery lustre at a fractured surface; carbon is present as carbide "cementite", Fe_3C. Being both hard and brittle, it has poor machining properties.

Wick-char Burning Test. A test for the carbon formation of kerosine. See **IP Burning Test.**

Wide Range Pressure Jet Burner. An *oil burner*, with a *turndown ratio*, qq.v., of up to 10 : 1. See **Spill Type Burner.**

W.I.F. Boiler. A sectional header *water-tube boiler*, q.v., it comprises straight solid drawn steel tubes inclined over the furnace at an angle of 15° to the horizontal, connecting the uptake header with the downtake header, each header being connected to the same water and steam drum. The wrought iron front (w.i.f.) today consists of steel. Designed and constructed by Babcock and Wilcox Ltd., the type has tended to be superseded by more modern bi-drum designs by the same makers. See **Boiler.**

Wigner Effect. The phenomenon in which an atom is displaced from its normal lattice position by neutron bombardment; the atom displaced may then come to rest in an interstitial position. The effect occurs in a graphite moderator and causes a change in shape of the graphite blocks. Moreover, if the atoms return subsequently to their normal positions the energy stored is given off in the form of heat.

Wigner Energy. The energy stored by the atoms of a moderator which have been displaced from their normal lattice positions by the *Wigner effect*, q.v. If the *moderator*, q.v., normally operates at a sufficiently high temperature the thermal agitation tends to cause the displaced atoms to return to their normal positions continuously;

this process is known as "annealing". With nuclear reactors which operate at lower temperatures it is, however, customary to carry out an annealing or Wigner energy release procedure periodically by raising the temperature above that used in normal operation.

Wind. Motion of the air caused by pressure differences. Wind direction determines where pollutants will go, while wind speed affects their dilution in the atmosphere. When wind speeds are high the area is well ventilated and pollutants do not accumulate. Most frequent wind speed in Britain is 16 ft/s at a height of 200 ft above ground level and 20 ft/s at 400 ft (See *Tables of Wind Direction and Forces over British Isles*, MOM 370, 2nd Ed, Air Ministry 1943).

Windscale Chemical Processing Plant. A U.K.A.E.A. plant located at Windscale, Cumberland, England, for the processing of irradiated fuel elements, i.e. burned-up nuclear reactor fuel elements. The elements are stored to allow the radioactive iodine to decay. After solution in acid the irradiated uranium passes through a complex chemical separation process which first separates the radioactive fission products. After the removal of the bulk of the fission products, the *plutonium* is separated from the *uranium*, qq.v. See **Uranium Enrichment.**

Windscale Incident. A major accident which occurred at the Windscale establishment of the U.K. Atomic Energy Authority on 10th October, 1957; a fire occurred inside the No. 1 reactor and volatile fission products were released to the atmosphere. The accident occurred during a routine Wigner release, i.e. the heating of graphite above its normal operating temperature to get rid of *Wigner energy*, q.v. Local overheating of the uranium fuel elements occurred, canning failed and exposed uranium oxidized. A filter in the ventilation stack arrested most of the strontium and caesium, but radioactive fission products escaped to atmosphere including Iodine 131 and Tellurium 132.

Winkler System. The gasification of coal utilizing a fluidized bed in a brick-lined chamber and an air/steam blast. Finely-divided fuel is fed into the grate by a screw; ash is removed from the grate by a plough. This system has been widely used in Germany on lignites.

Wiredrawing. The expansion of steam through an orifice where no work is done. The saturated steam at the lower pressure contains less heat per pound than the steam at the higher pressure; in this instance the release of a pound of steam makes heat available either to dry it or to superheat it. If the original steam is dry, its final temperature after expansion will be much higher than its saturated temperature due to a gain of superheat. For example, the *total heat*, q.v., of steam at 240 lb/in² and 470° F (243° C) is 1245 Btu/lb (692 Chu/lb). After passing through a reducing valve to give steam at 50 lb/in² the heat content of the saturated steam is 1180 Btu/lb (656 Chu/lb), making 65 Btu (36 Chu) available for superheat. Assuming a *specific heat*, q.v., of 0·5 this would give a final temperature of 428° F (220° C).

Wobbe Number. For any gas passing through a given orifice:

$$\text{Wobbe Number} = \frac{\text{c.v.}}{\sqrt{\text{sp. gr.}}}$$

where,   c.v. = calorific value of the gas;
       sp. gr. = specific gravity of the gas.
As the discharge of gas through an orifice in terms of Btu/h varies directly with the calorific value, but inversely with the square root of the specific gravity:

$$\text{Btu rate} = K \frac{\text{c.v.}}{\sqrt{\text{sp. gr.}}}$$

where K is a constant dependent on the orifice.

Wood. A *fuel*, q.v., consisting mainly of cellulose, a substance containing nearly 45 per cent of oxygen. The moisture content varies being about 25 per cent for freshly-felled timber and 15 per cent for air-dried timber. The *calorific value*, q.v., varies with the moisture content and the different amounts of oils and resins which different kinds of wood contain. A typical value for an oven-dried softwood is about 9000 Btu/lb, and for an oven-dried hardwood about 8500 Btu/lb. However with normal amounts of moisture calorific values of about 6000 Btu/lb are to be expected. Wood is a bulky fuel, being about half as dense as coal; to obtain the same amount of heat as any given weight of coal will yield, some four times the volume of wood must be burned. Wood is clean, readily ignites, burns with a long clean flame under correct combustion conditions, and leaves only a small amount of ash.

X

Xenon. Xe. An inert gas; at. no. 54, at. wt. 131·30; produced as a fission product in nuclear reactors. Xenon 135 has a very large neutron-capture cross-section and is one of the most important poisons in a reactor.

X-ray Diffraction. An accurate method of determining the details of the internal atomic structure of a substance; the planes of the atoms of crystals act as a diffraction grating to *X-rays*, q.v., which are scattered by them.

X-rays. Electromagnetic radiation, with a wavelength of less than 100 Å; produced by bombarding a metal target with fast electrons in an evacuated X-ray tube.

X-unit. The unit of wavelength for electromagnetic waves; equal to approximately 10^{-3} Ångström, or 10^{-11} centimetre. See **Spectroscopic Units.**

Y

Yorkshire Boiler. An early type of shell boiler, similar in design to the *Lancashire boiler*, q.v., save that the two internal furnace tubes were tilted upwards from front to rear, and the diameter of the tubes increased gradually from front to back. The purpose of this arrangement was to obtain uniform heat transmission.

Z

Zero Energy Reactor. An experimental *nuclear reactor*, q.v., operating at very low neutron flux and power level so that no forced cooling is required, and the fission product activity is so small that the fuel elements can be handled after use.

"Zig-zag" Filter. An *air filter*, q.v., in which the filtering medium of paper, fabric, glass fibre or plastic, is formed into a zig-zag shape over a wire framework. This type of filter incorporates a large filtration area in a comparatively small space. Collecting efficiences claimed depend on the filter medium, and vary from 45 to 95 per cent in the 0·1 to 5 micron particle size range.

Zirconium. Zr. A metallic *element*, q.v., at. no. 40, at. wt. 91·22; with a low neutron-capture cross-section. It has been considered as a possible canning material for fuel elements in nuclear reactors.

APPENDIX 1

Short Bibliography

The Efficient Use of Fuel (2nd Ed.) Ministry of Power, H.M.S.O., London, 1958.

Spiers, H.M. (Ed.)., *Technical Data on Fuel* 6th Ed., The British National Committee, World Power Conference, London, 1962.

Lyle, Sir Oliver, *The Efficient Use of Steam*, H.M.S.O., London, 1947.

Steam: Its Generation and Use, Babcock and Wilcox, New York.

Perry, John H., (Ed.) *Chemical Engineer's Handbook* (4th Ed.) McGraw-Hill Book Company, New York, 1963.

Thring, M. W., *The Science of Flames and Furnaces* (2nd Ed) Chapman and Hall, London, 1963.

Report of the Committee on Coal Derivatives (Cmnd 1120), H.M.S.O. London, 1960.

The New Stoker's Manual, National Industrial Fuel Efficiency Service, London.

Williams, J. N. *Boiler House Practice* (3rd Ed.), George Allen and Unwin, London, 1960.

Brame, J. S. S. and King, J. G. *Fuel*; *Solid, Liquid and Gaseous* (6th Ed.), Edward Arnold, London, 1967.

The Inorganic Constituents of Fuel (*Origin, Influence and Control*) Symposium held at the University of Melbourne, The Institute of Fuel, London, 1964.

Francis, W. *Coal: Its Formation and Composition*, Edward Arnold, London, 1954.

Francis, W. *Boiler House and Power Station Chemistry* (4th Ed.), Edward Arnold, London, 1962.

Francis, W. *Fuels and Fuel Technology*, Vols. 1 and 2, Pergamon Press, London, 1965.

Casci, C. (Ed.) *Fuels and New Propellants*, Pergamon Press, London, 1964.

Warren, F. A., *Rocket Propellants*, Reinhold Publishing Corporation, New York; Chapman and Hall, London, 1958.

Critchley, G. N., (Ed.) *The Future of Fuel Technology*, Pergamon Press, London, 1964.

Adams, P. J., *The Origin and Evolution of Coal* D.S.I.R., H.M.S.O., London, 1960.

Appendix 1

Handbook on Mechanical Stokers for Shell Boilers, National Coal Board, London, 1952.

Lowry, H. H., (Ed.) *Chemistry of Coal Utilization*, John Wiley & Sons, New York; Chapman and Hall, London, 1963.

Conference on Science in the Use of Coal, The Institute of Fuel, London, 1958.

Second Conference on Pulverized Fuel, The Institute of Fuel, London, 1957.

Conference on the Burning of Wood Waste, Auckland Industrial Development Division, D.S.I.R., Auckland, New Zealand, 1966.

Van Krevelen, D. W., *Coal*, Elsevier Publishing Coy, Amsterdam, 1961.

Williams, D. A. and Jones, G., *Liquid Fuels*, Pergamon Press and The Macmillan Company, London, 1963.

Sach, J. S., (Ed.) *Coal Tar Fuels; their Derivation, Properties and Application*, Association of Tar Distillers, London, 1960.

Modern Petroleum Technology, (3rd Ed.), The Institute of Petroleum London, 1962.

Third Conference on Liquid Fuels, The Institute of Fuel, London, 1966.

Noel, H. M., *Petroleum Refinery Manual*, Reinhold Publishing Corporation, New York; Chapman and Hall, London, 1959.

Gasification Processes, Proceedings of Joint Conference between The Institute of Fuel and The Institution of Gas Engineers, The Institute of Fuel, London, 1962.

Jay, K. *Nuclear Power, Today and Tomorrow*, Methuen, London, 1962.

Waste-Heat Recovery from Industrial Furnaces, Chapman and Hall, London, 1961. (Discussion relating to this Conference obtainable from The Institute of Fuel, London).

Badger, W. L. and Banchero, J. T. *Introduction to Chemical Engineering* McGraw-Hill, Book Company, New York, (1955).

Gilchrist, J. D. *Fuels and Refractories*, Pergamon Press, London; The MacMillan Company, New York, 1963.

Jackson, R. *Mechanical Equipment for Removing Grit and Dust from Gases*, British Coal Utilization Research Association, Leatherhead, England, 1963.

Hawksley, P. G. W., Badzioch, S. and Blackett, J. H. *Measurement of Solids in Flue Gases*, British Coal Utilization Research Association, Leatherhead, England, 1961.

Gilpin, A. *Control of Air Pollution*, Butterworths, London, 1963.

Report of the Committee on Air Pollution (Cmnd 9322), H.M.S.O., London, 1954.

The Hazards to Man of Nuclear and Allied Radiations (First Report, Cmnd. 9780, 1956; Second Report, Cmnd., 1225), Medical Research Council, H.M.S.O., London, 1960.

APPENDIX 2

Conversion of Some Common British and Other Units to Equivalent Values in the International System of Units (SI)

Length

1 in	= 2·54 cm
	= 25·4 mm
1 ft	= 0·3048 m
	= 30 48 cm
1 yd	= 0·9144 m
1 mile	= 1·60934 km
	= 1609·344 m
1 Å (ångström)	= 10^{-10} m

1 yd³ (cubic yard)

	= 0·764555 m³
	= 764·555 dm³
1 UK gal	= 4·5461 dm³
	= 4546·1 cm³
1 US gal	= 3·7854 dm³
	= 3785·4 cm³
1 litre (1901)	
	= 1000·028 cm³

Time

1 min	= 60 s
1 h	= 3·6 ks
1 day	= 86·4 ks
1 year	= 31·5 Ms

Velocity

1 in/s	= 2·54 cm/s
1 ft/s	= 0·3048 m/s
1 mile/h	= 0·44704 m/s
1 UK knot	= 0·51478 m/s

Area

1 in² (square inch)	
	= 6·4516 cm²
	= 645.16 mm²
1 ft² (square foot)	
	= 929·030 cm²
	= 0·092903 m²
1 yd² (square yard)	
	= 0·836127 m²
1 acre	= 4046·9 m²
1 mile²	= 2·58999 km²

Moment of inertia

1 lb ft² = 0·0421401 kg m²

Momentum

1 lb ft/s = 0·13826 kg m/s

Angular momentum

1 lb ft²/s = 0·042140 kg m²/s

Acceleration

1 ft/s² (foot per second per second) = 0·3048 m/s²

Volume

1 in³ (cubic inch)	
	= 16·3871 cm³
1 ft³ (cubic foot)	
	= 28·3168 dm³
	= 0·02832 m³

Mass

1 gr (grain)	
	= 64·7989 mg
1 oz	= 28·3495 g
1 lb	= 0·45359237 kg

263

1 cwt = 50·8023 kg
1 UK ton (long)
  = 1016·05 kg
  = 1·01605 tonne
    (1000 kg)
1 slug = 14·5939 kg

Mass per unit area
1 oz/ft² = 305·152 g/m²
1 oz/yd² = 33·9057 g/m²
1 lb/in² = 70·3070 g/cm²
  = 703·07 kg/m²
1 lb/ft² = 4·88243 kg/m²
1 UK ton/sq mile
  = 392·298 kg/km²

Mass rate of flow (*mass/time*)
1 lb/s = 0·453592 kg/s
1 lb/h = 0·12600 g/s
1 UK ton/h = 0·28224 kg/s

Volume rate of flow (*volume/time*)
1 ft³/s (1 cusec)
  = 0·0283168 m³/s
  = 28·3168 dm³/s
1 ft³/h = 7·8658 cm³/s
1 UK gal/h = 1·2628 cm³/s
1 US gal/h = 1·0515 cm³/s

Density (*mass/volume*)
1 lb/in³ = 27·6799 g/cm³
1 lb/ft³ = 16·0185 kg/m³
1 UK ton/yd³ = 1328·94 kg/m³
  = 1·32894
    tonne/m³
1 slug/ft³ = 515·379 kg/m³
1 lb/UK gal = 99·776 kg/m³
1 lb/US gal = 119·83 kg/m³

Concentration (*mass/volume*)
1 gr/ft³   = 2·28835 g/m³
1 gr/UK gal = 14·2538 g/m³

Force
1 pdl (poundal)
  = 0·138255 N
    (newton)
1 lbf = 4·44822 N
1 kgf = 9·8067 N
1 UK tonf = 9·96402 kN
1 dyne = 10^{-5} N

Pressure
1 mm Hg = 133·32 N/m²
1 in Hg = 3·3864 kN/m²
1 in H₂O = 249·089 N/m²
1 ft H₂O = 2·9891 kN/m²
1 pdl/ft² = 1·48816 N/m²
1 lbf/in² = 6·8948 kN/m²
1 lbf/ft² = 47·8803 N/m²
1 UK tonf/in²
  = 15·4443 MN/m²
1 UK tonf/ft²
  = 107·252 kN/m²
1 atm (std) = 101·325 kN/m²
1 bar = 10^5 N/m²

Viscosity (*dynamic*)
1 lbf s/ft² = 47·8803 N s/m²
1 pdl s/ft² = 1·48816 N s/m²
  = 1488·16
    centipoise, cP
1 lb/ft s = 1·48816 N s/m²
  = 1488·16
    centipoise, cP
1 lb/ft h = 0·41338 MN s/m²
1 slug/ft s = 47·8803 kg/m s
1 poise, P = 10^{-1} N s/m²
1 centipoise, cP
  = 10^{-3} N s/m²

Viscosity (kinematic)
1 in²/s   = 6·4516 cm²/s
          = 645·16
          centistokes, cSt
1 ft²/s   = 0·0929030 m²/s
1 ft²/h   = 0·25806 cm²/s
          = 25·8064
          centistokes, cSt
1 stokes, St = 10^{-4} m²/s
1 centistokes, cSt
          = 10^{-6} m²/s

Surface tension
1 dyn/cm = 10^{-3} N/m
1 erg/cm² = 10^{-3} J/m²

Energy (work, heat)
1 therm   = 105·506 MJ
1 thermie = 4·1855 MJ
1 kWh   = 3·6 MJ
1 Btu   − 1·05506 kJ
1 ft lbf   = 1·35582 J
1 ft pdl   = 0·0421401 J
1 cal (I.T.) = 4·1868 J
1 cal (15°) = 4·1855 J
1 thermochemical calorie
          = 4·184 J
1 erg   = 10^{-7} J
1 hp h   = 2·6845 MJ

Power
1 hp (metric) = 735·50 W
1 hp (British) = 745·700 W
1 erg/s   = 10^{-7} W
1 ft lbf/s   = 1·3558 W

Heat flow rate
1 Btu/h = 0·293071 W
1 cal/s = 4·1868 W

1 kcal/h = 1·163 W
1 ton of refrigeration
          = 3516·9 W

Calorific value
1 therm/UK gal
          = 23·2080 MJ/dm³
1 Btu/ft³ = 37·2589 kJ/m³
1 Btu/lb = 2·326 kJ/kg

Temperature difference
1 degF (degR)
          = 5/9 deg C (deg K)

Thermal capacity
1 Btu/lb degF
          = 4·1868 kJ/kg degK
1 Btu/ft³ degF
          67·0661 kJ/m³ degK

Heat release
1 Btu/ft³ h
          = 1·03497 × 10^{-5} W/cm³

Thermal conductance
1 Btu/ft² h degF
          = 5·67826 W/m² degK

Thermal conductivity
1 Btu/ft h degF
          = 1·7307 W/m degK

Illumination
1 lm/ft² = 1 foot-candle
          = 10·7639 lx (lm/m²)

Luminance
1 cd/ft² = 10·7639 cd/m²

Appendix 2

Useful Additional Data

Length

1 micron (μ)	10^{-4} cm
	$= 3.9 \times 10^{-5}$ in
	$= 10^4$ ångström
1 ångström (Å)	$= 10^{-8}$ cm
	$= 10^{-10}$ m
1 millimetre	$= 1000$ microns
1 centimetre	$= 0.3937$ in
	$= 0.03281$ ft
	$= 10$ mm
1 decimetre	$= 10$ cm
1 metre	$= 1.094$ yd
	$= 3.281$ ft
	$= 39.370$ in
	$= 10$ dm
1 kilometre	$= 0.6214$ mile
	$= 1000$ m

Area

1 cm²	$= 0.155$ in²
1 m²	$= 10.764$ ft²
	$= 1.196$ yd²
1 km²	$= 0.3861$ sq mile
1 mile²	$= 640$ acres
1 acre	$= 4840$ yd²

Volume

1 ft³	$= 28.32$ litres
	$= 6.229$ gal (water at 62° F, 16.7° C)
1 cm³	$= 0.061$ in³
1 dm³	$= 0.0353147$ ft³
1 m³	$= 35.315$ ft³
	$= 1.308$ yd³
1 UK gallon	$= 1.2$ US gallon
1 US gallon	$= 0.833$ UK gallon

1 cusec	$= 1$ ft³/s
1 litre	$= 0.03532$ ft³
	$= 0.220$ UK gallons
	$= 0.264$ US gallons
	$= 1000.028$ cm³
1 millilitre	$= 1$ cm³

Velocity

| 1 km/h | $= 0.621371$ mile/h |
| 1 m/s | $= 3.28084$ ft/s |

Acceleration

1 m/s² $= 3.281$ ft/s²

Mass

1 lb	$= 453.6$ g
	$= 7000$ grains (gr)
1 gr (grain)	$= 0.0648$ g
1 g (gramme)	$= 15.432$ gr
	$= 0.0022$ lb
1 kg (kilogramme)	$= 2.205$ lb
1 long ton (UK)	$= 2240$ lb
	$= 1.016$ metric ton (tonne)
	$= 1.120$ short tons
1 metric ton (tonne)	$= 2204.620$ lb
	$= 0.984$ long ton
	$= 1.102$ short tons
	$= 1000$ kg

1 short ton (US)
        = 2000 lb
        = 0·907 metric
          ton
        = 0·893 long ton
1 slug   = 32·1740 lb

Force
1 N = 0·224809 lbf
    = 7·23301 pdl

Pressure
1 N/m² = 0·000145038 lbf/in²
1 bar   = 10⁶ dyn/cm²
       = 750·1 mm Hg

Density
1 g/cm³ = 62·43 lb/ft³
1 kg/m³ = 0·062428 lb/ft³

Viscosity, dynamic
1 N s/m² = 0·0208854 lbf s/ft²

Viscosity, kinematic
1 m²/s = 10·7639 ft²/s

Concentration
1 gr/ft³ = 2·29 g/m³
1 g/m³ = 0·44 gr/ft³

Energy
1 J = 0·737562 ft lbf
1 kJ = 0·277778 Wh

Heat Units

Amount of water heated	Range through which heated	Name of unit
1 pound	One degF	British thermal unit (Btu)
1 pound	One degC	Centigrade (or Celsius) heat unit (Chu)
1 gramme	One degC	calorie (cal)
1 kilogramme	One degC	kilogramme calorie (kcal)

1 calorie = 4·1868 joule (J)
1 Btu   = 252 cal
         = 0·252 kcal
         = 0·556 Chu
         = 1055·06 J
1 therm = 100,000 Btu
         = 55,555 Chu
         = 25,200 kcal
1 Chu   = 1·8 Btu
         = 454 cal
1 megawatt (MW)
         = 1 million watts

1 kilowatt hour (kWh)
         = 1000 watt-hours
         = 3412 Btu
         = 1896 Chu
         = 3600 kJ

1 watt (W) = 1 joule per
          second
         = 10⁷ erg/s

1 joule (J) = 10⁷ erg
         = 10⁷ dyn cm

Temperature Conversions

$T°$ F is equivalent to $5/9 (T - 32)°$ C
$T°$ C is equivalent to $(9/5T + 32)°$ F
$T°$ F is equivalent to $(T + 460)°$ R
$T°$ C is equivalent to $(T + 273)°$ K

Mathematical Constants

$\pi = 3\cdot1416$
$e = 2\cdot7183$

Physical Constants

Acceleration due to standard gravity	$= 980\cdot665$ centimetres per second squared (cm/s^2)
	$= 32\cdot174$ feet per second squared (ft/s^2)
Normal atmospheric pressure in inches of mercury (in Hg) at $32°$ F ($0°$ C) under standard gravity	$= 29\cdot92$ in Hg
Normal atmospheric pressure in millimetres of mercury (mm Hg) at $32°$ F ($0°$ C) under standard gravity	$= 760\cdot0$ mm Hg
Normal atmospheric pressure, in pounds per square inch, under standard gravity	$= 14\cdot70$ lb/in^2
Normal temperature and pressure (n.t.p.)	$= 0°$ C and 760 mm Hg
Standard temperature and pressure (s.t.p.)	$= 0°$ C and 760 mm Hg
Density of water at $20°$ C in pounds per gallon	$= 10\cdot0$ lb/gal
Absolute zero in degrees celsius	$= -273\cdot2$
Absolute zero in degrees fahrenheit	$= -459\cdot7$
Atomic weight of carbon	$= 12\cdot01$
Atomic weight of oxygen	$= 16\cdot00$
Atomic weight of hydrogen	$= 1\cdot008$
Atomic weight of nitrogen	$= 14\cdot01$
Atomic weight of sulphur	$= 32\cdot06$
Density of dry air at n.t.p. in pounds per cubic foot	$= 0\cdot0807$ lb/ft^3

Miscellaneous Conversion Factors

To convert	*Multiply by*
Feet per second to miles per hour	0·6818
Miles per hour to feet per second	1·4667
Miles per hour to centimetres per second	44·704
Feet of water to inches of mercury	0·8843 (at 20° C)
Feet of water to pounds per square inch	0·4327 (at 20° C)
Metres per second to feet per second	3·281
Millibars to atmospheres	$9·869 \times 10^{-4}$

Metric Prefixes

deka (da) = $\times$ 10	deci (d)	= $\times$ 10^{-1}
hecto (h) = $\times$ 10^2	centi (c)	= $\times$ 10^{-2}
kilo (k) = $\times$ 10^3	milli (m)	= $\times$ 10^{-3}
mega (M) = $\times$ 10^6	micro (μ)	= $\times$ 10^{-6}
giga (G) = $\times$ 10^9	nano (n)	= $\times$ 10^{-9}
tera (T) = $\times$ 10^{12}	pico (p)	= $\times$ 10^{-12}

APPENDIX 3

Tables of Conversion Factors for Units used in Atmospheric Pollution Calculations

1. *Sulphur Dioxide Concentrations*

From	To (*Multiplying Factors*)					
	p.p.h.m. v/v	mg/100 m^3	mg/m^3	μg/m^3	mg SO$_3$/ 100 cm^2 day	Grains ft^3
p.p.h.m.	1	2·86	$2·86 \times 10^{-2}$	28·6	$2·6 \times 10^{-1}$	$1·3 \times 10^{-5}$
mg/100m^3	0·35*	1	0·01	10	$8·9 \times 10^{-2}$	$4·4 \times 10^{-6}$
mg/m^3	35	100	1	10^3	8·9	$4·4 \times 10^{-4}$
μg/m^3	0·035	0·10	0·001	1	$8·9 \times 10^{-3}$	$4·4 \times 10^{-7}$
mg SO$_3$/ 100 cm^2 day	3·9	11	0·11	111·5†	1	$4·9 \times 10^{-5}$
Grains/ ft^3	$8·0 \times 10^4$	$2·3 \times 10^5$	$2·3 \times 10^3$	$2·3 \times 10^6$	$2·1 \times 10^4$	1

* Report of Committee on Air Pollution, 1954.
† D.S.I.R. Thirty-first report on *The Investigation of Atmospheric Pollution*, 1960.

2. *Ground Level Dust Depositions*

From	To (*Multiplying factor*)		
	Ton/mile^2 month	g/100 m^2 month	mg/m^2 day
Ton/mile^2 month	1	39	13
g/100 month^2	0·026	1	0·33
mg/m^2 day	0·077	3	1

APPENDIX 4

Temperature Conversion Table

To convert any temperature in Centigrade or Fahrenheit degrees to the other, select the figure to be converted in the centre column, and if converting from Centigrade to Fahrenheit, read off to the right-hand column; if converting from Fahrenheit to Centigrade, read off to the left-hand column.

Examples:— 1400° C = 2552° F
390° F = 199° C

C		F	C		F	C		F
−17·8	0	32	1·67	35	95·0	21·1	70	158·0
−17·2	1	33·8	2·22	36	96·8	21·7	71	159·8
−16·7	2	35·6	2·78	37	98·6	22·2	72	161·6
−16·1	3	37·4	3·33	38	100·4	22·8	73	163·4
−15·6	4	39·2	3·89	39	102·2	23·3	74	165·2
−15·0	5	41·0	4·44	40	104·0	23·9	75	167·0
−14·4	6	42·8	5·00	41	105·8	24·4	76	168·8
−13·9	7	44·6	5·56	42	107·6	25·0	77	170·6
−13·3	8	46·4	6·11	43	109·4	25·6	78	172·4
−12·8	9	48·2	6·67	44	111·2	26·1	79	174·2
−12·2	10	50·0	7·22	45	113·0	26·7	80	176·0
−11·7	11	51·8	7·78	46	114·8	27·2	81	177·8
−11·1	12	53·6	8·33	47	116·6	27·8	82	179·6
−10·6	13	55·4	8·89	48	118·4	28·3	83	181·4
−10·0	14	57·2	9·44	49	120·2	28·9	84	183·2
− 9·44	15	59.0	10·0	50	122·0	29·4	85	185·0
− 8·89	16	60·8	10·6	51	123·8	30·0	86	186·8
− 8·33	17	62·6	11·1	52	125·6	30·6	87	188·6
− 7·78	18	64·4	11·7	53	127·4	31·1	88	190·4
− 7·22	19	66·2	12·2	54	129·2	31·7	89	192·2
− 6·67	20	68·0	12·8	55	131·0	32·2	90	194·0
− 6·11	21	69·8	13·3	56	132·8	32·8	91	195·8
− 5·56	22	71·6	13·9	57	134·6	33·3	92	197·6
− 5·00	23	73·4	14·4	58	136·4	33·9	93	199·4
− 4·44	24	75·2	15·0	59	138·2	34·4	94	201·2
− 3·89	25	77·0	15·6	60	140·0	35·0	95	203·0
− 3·33	26	78·8	16·1	61	141·8	35·6	96	204·8
− 2·78	27	80·6	16·7	62	143·6	36·1	97	206·6
− 2·22	28	82·4	17·2	63	145·4	36·7	98	208·4
− 1·67	29	84·2	17·8	64	147·2	37·2	99	210·2
− 1·11	30	86·0	18·3	65	149·0	38	100	212
− 0·56	31	87·8	18·9	66	150·8	43	110	230
0	32	89·6	19·4	67	152·6	49	120	248
0·56	33	91·4	20·0	68	154·4	54	130	266
1·11	34	93·2	20·6	69	156·2	60	140	284

Appendix 4

C		F	C		F	C		F
66	**150**	302	338	**640**	1184	616	**1140**	2084
71	**160**	320	343	**650**	1202	621	**1150**	2102
77	**170**	338	349	**660**	1220	627	**1160**	2120
82	**180**	356	354	**670**	1238	632	**1170**	2138
88	**190**	374	360	**680**	1256	638	**1180**	2156
93	**200**	392	366	**690**	1274	643	**1190**	2174
99	**210**	410	371	**700**	1292	649	**1200**	2192
100	**212**	414	377	**710**	1310	654	**1210**	2210
104	**220**	428	382	**720**	1328	660	**1220**	2228
110	**230**	446	388	**730**	1346	666	**1230**	2246
116	**240**	464	393	**740**	1364	671	**1240**	2264
121	**250**	482	399	**750**	1382	677	**1250**	2282
127	**260**	500	404	**760**	1400	682	**1260**	2300
132	**270**	518	410	**770**	1418	688	**1270**	2318
138	**280**	536	416	**780**	1436	693	**1280**	2336
143	**290**	554	421	**790**	1454	699	**1290**	2354
149	**300**	572	427	**800**	1472	704	**1300**	2372
154	**310**	590	432	**810**	1490	710	**1310**	2390
160	**320**	608	438	**820**	1508	716	**1320**	2408
166	**330**	626	443	**830**	1526	721	**1330**	2426
171	**340**	644	449	**840**	1544	727	**1340**	2444
177	**350**	662	454	**850**	1562	732	**1350**	2462
182	**360**	680	460	**860**	1580	738	**1360**	2480
188	**370**	698	466	**870**	1598	743	**1370**	2498
193	**380**	716	471	**880**	1616	749	**1380**	2516
199	**390**	734	477	**890**	1634	754	**1390**	2534
204	**400**	752	482	**900**	1652	760	**1400**	2552
210	**410**	770	488	**910**	1670	766	**1410**	2570
216	**420**	788	493	**920**	1688	771	**1420**	2588
221	**430**	806	499	**930**	1706	777	**1430**	2606
227	**440**	824	504	**940**	1724	782	**1440**	2624
232	**450**	842	510	**950**	1742	788	**1450**	2642
238	**460**	860	516	**960**	1760	793	**1460**	2660
243	**470**	878	521	**970**	1778	799	**1470**	2678
249	**480**	896	527	**980**	1796	804	**1480**	2696
254	**490**	914	532	**990**	1814	810	**1490**	2714
260	**500**	932	538	**1000**	1832	816	**1500**	2732
266	**510**	950	543	**1010**	1850	821	**1510**	2750
271	**520**	968	549	**1020**	1868	827	**1520**	2768
277	**530**	986	554	**1030**	1886	832	**1530**	2786
282	**540**	1004	560	**1040**	1904	838	**1540**	2804
288	**550**	1022	566	**1050**	1922	843	**1550**	2822
293	**560**	1040	571	**1060**	1940	849	**1560**	2840
299	**570**	1058	577	**1070**	1958	854	**1570**	2858
304	**580**	1076	582	**1080**	1976	860	**1580**	2876
310	**590**	1094	588	**1090**	1994	866	**1590**	2894
316	**600**	1112	593	**1100**	2012	871	**1600**	2912
321	**610**	1130	599	**1110**	2030	877	**1610**	2930
327	**620**	1148	604	**1120**	2048	882	**1620**	2948
332	**630**	1166	610	**1130**	2066	888	**1630**	2966

C	F		C	F		C	F	
893	**1640**	2984	1154	**2110**	3830	1416	**2580**	4667
899	**1650**	3002	1160	**2120**	3848	1421	**2590**	4694
904	**1660**	3020	1166	**2130**	3866	1427	**2600**	4712
910	**1670**	3038	1171	**2140**	3884	1432	**2610**	4730
916	**1680**	3056	1177	**2150**	3902	1438	**2620**	4748
921	**1690**	3074	1182	**2160**	3920	1443	**2630**	4766
927	**1700**	3092	1188	**2170**	3938	1449	**2640**	4784
932	**1710**	3110	1193	**2180**	3956	1454	**2650**	4802
938	**1720**	3128	1199	**2190**	3974	1460	**2660**	4820
943	**1730**	3146	1204	**2200**	3992	1466	**2670**	4838
949	**1740**	3164	1210	**2210**	4010	1471	**2680**	4856
954	**1750**	3182	1216	**2220**	4028	1477	**2690**	4874
960	**1760**	3200	1221	**2230**	4046	1482	**2700**	4892
966	**1770**	3218	1227	**2240**	4064	1488	**2710**	4910
971	**1780**	3236	1232	**2250**	4082	1493	**2720**	4928
977	**1790**	3254	1238	**2260**	4100	1499	**2730**	4946
982	**1800**	3272	1243	**2270**	4118	1504	**2740**	4964
988	**1810**	3290	1249	**2280**	4136	1510	**2750**	4982
993	**1820**	3308	1254	**2290**	4154	1516	**2760**	5000
999	**1830**	3326	1260	**2300**	4172	1521	**2770**	5018
1004	**1840**	3344	1266	**2310**	4190	1527	**2780**	5036
1010	**1850**	3362	1271	**2320**	4208	1532	**2790**	5054
1016	**1860**	3380	1277	**2330**	4226	1538	**2800**	5072
1021	**1870**	3398	1282	**2340**	4244	1543	**2810**	5090
1027	**1880**	3416	1288	**2350**	4262	1549	**2820**	5108
1032	**1890**	3434	1293	**2360**	4280	1554	**2830**	5126
1038	**1900**	3452	1299	**2370**	4298	1560	**2840**	5144
1043	**1910**	3470	1304	**2380**	4316	1566	**2850**	5162
1049	**1920**	3488	1310	**2390**	4334	1571	**2860**	5180
1054	**1930**	3506	1316	**2400**	4352	1577	**2870**	5198
1060	**1940**	3524	1321	**2410**	4370	1582	**2880**	5216
1066	**1950**	3542	1327	**2420**	4388	1588	**2890**	5234
1071	**1960**	3560	1332	**2430**	4406	1593	**2900**	5252
1077	**1970**	3578	1338	**2440**	4424	1599	**2910**	5270
1082	**1980**	3596	1343	**2450**	4442	1604	**2920**	5288
1088	**1990**	3614	1349	**2460**	4460	1610	**2930**	5306
1093	**2000**	3632	1354	**2470**	4478	1616	**2940**	5324
1099	**2010**	3650	1360	**2480**	4496	1621	**2950**	5342
1104	**2020**	3668	1366	**2490**	4514	1627	**2960**	5360
1110	**2030**	3686	1371	**2500**	4532	1632	**2970**	5378
1116	**2040**	3704	1377	**2510**	4550	1638	**2980**	5396
1121	**2050**	3722	1382	**2520**	4568	1643	**2990**	5414
1127	**2060**	3740	1388	**2530**	4586	1649	**3000**	5432
1132	**2070**	3758	1393	**2540**	4604	1705	**3100**	5619
1138	**2080**	3776	1399	**2550**	4622	1760	**3200**	5722
1143	**2090**	3794	1404	**2560**	4640	1816	**3300**	5972
1149	**2100**	3812	1410	**2570**	4658	1871	**3400**	6152